EVOLVING GOD

EVOLVING God

A PROVOCATIVE VIEW
OF THE ORIGINS OF RELIGION

Barbara J. King

DOUBLEDAY

NEW YORK   LONDON   TORONTO   SYDNEY   AUCKLAND

PUBLISHED BY DOUBLEDAY

Book design by Caroline Cunningham

Cataloging-in-Publication Data is on file with the Library of Congress

ISBN: 978-0-385-51104-9

PRINTED IN THE UNITED STATES OF AMERICA

1   3   5   7   9   10   8   6   4   2

First Edition

For those I love, of several species,

Who sustain me every day

And who know:

"It ain't no sin to be glad you're alive"

—BRUCE SPRINGSTEEN

CONTENTS

CONTENTS

ONE

Apes to Angels

WE HUMANS CRAVE emotional connection with others. This deep desire to connect can be explained by the long evolutionary history we shared with other primates, the monkeys and apes. At the same time, it explains why humans evolved to become the spiritual ape—the ape that grew a large brain, the ape that stood up, the ape that first created art, but, above all, the ape that evolved God.

A focus on emotional connection is an exciting way to view human prehistory, but it is not the traditional way. Millions of years of human evolution are most often recounted as a series of changes in the skeletons, artifacts, and big, flashy, attention-grabbing behaviors of our ancestors. Medium-size skulls with forward-jutting jaws morph into skulls with high foreheads, large enough to house a neuron-packed human brain. Bones of the leg lengthen and shape-shift over time, so that a foot with apelike curved toes becomes a foot that imprints the sand just the way yours and mine do as we stroll along the surf. Crudely modified tools made of rough stone develop gradually into objects of antler and bone, delicately fashioned and as much symbolic as utilitarian. Caves, at first refuges for Neandertal hunters seeking shelter from hungry bears and other carnivores, become colorful art galleries when *Homo sapiens* begins to paint the walls with magnificent images of the animals they hunt.

Stones, bones, and "big" behaviors like tool-making and cave-painting do change over time as our ancestors evolve, and much of what we can learn about these transformations is enlightening. But the most profound, indeed the most stirring transformations in the evolutionary history of *Homo sapiens* involve what does not fossilize and what is only sometimes made tangible: belongingness.

Belongingness is mattering to someone who matters to you. It's about getting positive feelings from our relationships. It's what you and I work to maintain (or what we wish for) with family and friends, and perhaps also with colleagues or people in our community; for some of us, it extends to animals as well (*other* animals, for we humans are first and foremost animals). Relating emotionally to others shapes the very quality of our lives.

Belongingness, then, is a useful shorthand term for the undeniable reality that humans of all ages, in all societies, thrive in relation to others. That humans crave emotional connection is obvious in some respects. Most of us marry and live in families, configured either as parents (or a single parent) living with children or, more commonly worldwide, as multiple generations living together in extended family groups. We do things, both spiritual and secular, and by choice as well as necessity, in groups of relatives, friends, and associates. We write great literature and make great art based on the deepest emotions for those we love, or pine for, or grieve for.

Who can linger over a superbly crafted love poem and doubt the depth of human yearning for belongingness? We *feel*, rather than merely read or hear, Emily Dickinson's poem "Compensation":

For each extatic instant
We must an anguish pay

In keen and quivering ratio
To the ecstasy.
For each beloved hour
Sharp pittances of years,
Bitter contested farthings
And coffers heaped with tears.

For one reader, these words might conjure up two lovers separated, by death or by mere circumstance, after a too-fleeting time together, an image accompanied by a feeling of searing loss. For another reader, they might bring to mind what happens when a cherished child not only grows up but grows apart, a thought coupled with a bittersweet mingling of pride and regret at being the center of her universe no more.

Emerging from the emotional depths of this poem, a reader might wonder what new can be said about human belongingness that might shed light on the evolution of the human religious imagination. Compelling questions can guide us here.

WEAVING A STORY

How did humans go from craving belongingness to relating in profound and deep ways to God, gods, or spirits? How did an engagement with the sacred that is wholly unique to humans emerge from a desire for belongingness that is common to monkeys, apes, extinct human ancestors, and humans of today? These seem to me the most vital questions, and they will act as my touchstone as I weave two thick strands of information together into an evolutionary account of the prehistory of belongingness.

For two and a half decades, at work in zoos and research centers and in the African bush, I have observed, filmed, and interpreted the behavior of monkeys and apes. The social and emotional behavior of these close relatives of ours never fails to fascinate in its own right. In long-term study of particular social groups, any keen observer comes to

recognize bitter rivalries, deep friendships, and enduring family ties—
and becomes convinced that the animals, too, recognize them and act
accordingly.

Like most anthropologists, however, I have been motivated ultimately by the wish to understand better the behavior of my own
species. Coupling my own research with analysis of the behavior of our
humanlike extinct ancestors in Africa, Asia, and Europe—as studied
by other scholars—has allowed me to grasp something about just how
we humans evolved. I am especially fascinated with the evolutionary
history of empathy; of meaning-making; of rule-following; of imagination; and of consciousness. In what ways do monkeys and apes today
express behaviors related to these aspects of emotional and cognitive
life? How can we best seek evidence of these in our extinct ancestors?
Can we uncover traces of our emotional prehistory in the remains,
both physical and cultural, of the Neandertals and related groups? If
so, how do these traces speak to us across the millennia about the development of religion?

These questions emerge from my own experience as an observer of
primates, a writer, and a student of others' anthropological research—
and indeed from my long-standing tendency to be attracted to the "big
questions" of biological anthropology. Yet no book that purports to explain something meaningful about religion can spring entirely from a
single discipline. Though biological anthropology is the most appropriate field in which to ground our inquiry, it's necessary to adopt a
broad perspective.

A second set of issues beckons us further into the labyrinth that
must be negotiated in any study of religion. What *is* religion? What is
the relationship—both in the present and in the past—between religious belief and religious practice? That is, must religion be defined as
a set of beliefs, or can it be something different? How do theologians
and other religious thinkers portray the relationship between faith and
practice? Can understanding this relationship lead us to a different
take on the findings from the first set of questions, those about the prehistory of religion?

The challenge is to weave together two discrete strands: the development of the religious imagination throughout prehistory, and the phenomenon of religion itself. These two threads, each with a panoply of attendant questions, seem to lead in a dizzying variety of directions. In the following chapters, I shall draw the threads together into a coherent story. Along the way, I will compare and contrast my views with those of other writers who speculate about the origins of religion. In what ways are these theorists on the right track, and in what ways do they miss critical pieces of the puzzle?

For now, the essence of my argument can be summarized in three key points:

A fundamental characteristic of all primates, the need for belongingness is most elaborated in the African apes, our closest living relatives. Though we did not descend from chimpanzees or gorillas, we share with them a common ancestor. The everyday social behavior of this apelike ancestor laid a foundation for the evolution of religion that was to come much later, a foundation that can be reconstructed from knowledge of what today's apes do.

Drawing on my own years of up-close-and-personal encounters with chimpanzees and gorillas, I discuss in Chapter 2 the early precursors to religion—empathy, meaning-making, rule-following, and imagination—and how these relate to the issue of ape consciousness. I am convinced that apes are highly sensitive and tuned in to one another starting in infancy, when a baby begins to negotiate with its mother about its needs. More than most other mammals, ape infants are born into a highly social world, a web of emotional interactions among relatives and other social partners. Research on animals like dolphins and elephants may someday challenge this conclusion, but it seems clear at least that the way two apes respond to each other sensitively and contingently is of different quality than what happens when two wolves, say, or two domestic cats, circle each other and adjust to each other's snarls, or lunges, in a well-honed, highly instinctual dance. It even seems different from the learned behaviors of other primates, like monkeys. The apes' finely tuned responses to each other are rooted in belongingness, in the emotionality toward others that stems from their being so keenly dependent on their mothers and other relatives from birth onward.

Second, profound changes in emotional relating occurred as our human ancestors' lives diverged from those of the apelike ancestors. In Chapters 3 through 6, I focus on the origins of the human religious imagination in the span of time bounded, on the one end, by the divergence of hominids (human ancestors) from the ape lineage about 6 million or 7 million years ago, and on the other by the beginning of farming and settled communities around 10,000 years ago. Admittedly, we can glean almost nothing concrete about emotional connectedness as far back as 7 million years (though we can continue to use modern-day apes as models, and speculate in useful ways). After 3 million years ago, the record of material culture—fossilized artifacts and other concrete products of hominid behavior—begins. At that point, tangible clues help us assess the changes that take place in empathy, meaning-making, rule-following, imagination, and consciousness, and, indeed, in the pattern of nurturing and caring that lays the foundation for all of these.

After all, it is not the stones and bones, the technology and art, that deserve top billing in our prehistory; it is material culture's emotional backstory that does. Throughout the millennia, hominid mothers nurtured their children; siblings played with each other and with their friends; adults shifted alliances, supporting first this friend, then another, against a rival. The emotional dependency of ape infants on their mothers and other relatives only deepened and lengthened as the human lineage began to evolve, a fact with cascading consequences for the hominids' whole lives.

The archaeologist Steven Mithen rescues Neandertals, for instance, from the caveman-dragging-cavewoman-by-the-hair stereotype by acknowledging this rich inner life; he writes of "intensely emotional beings: happy Neanderthals, sad Neanderthals, angry Neanderthals, disgusted Neanderthals, envious Neanderthals, guilty Neanderthals, grief-stricken Neanderthals, and Neanderthals in love."[1] While I embrace Mithen's sensibility, I would have put the statement a bit differently: "Neandertals making each other happy, Neandertals making each other sad . . ." Emotions, before, after, and during the Neandertal period, are created when individuals act together and make meaning together, starting in infancy. The excitement in understanding human evolution is centered in tracing this mutual creativity and meaning-making, indeed in tracing the evolution of belongingness.

Third, the hominid need for belongingness rippled out, eventually expanding into a wholly new realm. In tandem with, and in part driven by, changes in the natural environment, in the hominid brain, and most important, in caregiving practices, something new emerged that went beyond empathy, rule-following, and imagination within the family and immediate group, and that went beyond consciousness expressed through action and meaning-making in the here and now. As I explain in Chapters 6 and 7, language and culture became more complex as symbols and ritual practices began to play a more central role in how hominids made sense of their world. An earthly need for belongingness led to the human religious imagination and thus to the otherworldly realm of relating with God, gods, and spirits.

From the building blocks we find in apelike ancestors emerged the soulful need to pray to gods, to praise God with hymns, to shake in terror before the power of invisible spirits, to fear for one's life at the hands of the unknown or to feel bathed in all-enveloping love from the heavens. To express in straightforward language the profound depth of this human emotional connection to the sacred is a challenge. The inaccessibility to language of the sacred experience mirrors what Martin Buber writes about when he describes human relating with God: it "is wrapped in a cloud but reveals itself, it lacks but creates language. We hear no You and yet we feel addressed; we answer—creating, thinking, acting: with our being we speak the basic word, unable to say You with our mouth."[2]

Buber's *I and Thou* is a wonderful (in the word's literal sense) lead-in to understanding my thesis. Buber says that "all actual life is encounter," that "in the beginning is the relation," that "man becomes an I through a You."[3] This is so and has been so for a very long time in our prehistory. What's so beautiful and compelling about the human religious imagination in all its ineffable relating is how it emerges from its evolutionary precursors and yet completely transfigures them.

In highlighting this critical balance between evolutionary continuity and evolutionary transformation, I want to be crystal clear about the role of belongingness in the origins of religion. I see belongingness as one aspect of religiousness—an aspect so essential that the human religious imagination could not have evolved without it. In scientific lingo, belongingness is a *necessary condition* for the evolution of religion. Over the course of prehistory, belongingness was transformed from a basic emotional relating between individuals to a deeper relating, one that had the potential to become *transcendent*, between people and supernatural beings or forces.

My focus on belongingness distinguishes my perspective from the dominant one today. In our age of high-tech science, when gene sequencing and brain-mapping reign supreme, it is little surprise to find that the most popular theories of the origin of religion center around properties of genes and brains. Specific genetic-biochemical profiles and inherited brain "modules" devoted to the expression of religion animate these theories. While something can be learned from such scenarios, they are sterile to the degree that they fail to grasp the significance of what matters most: people deeply and emotionally engaged with others of their kind, and eventually with the sacred.

That social interactions played a central role in the origins of religion is not, of course, an original insight. Such an emphasis may no longer be favored, but at least since the time of the pioneering sociologist Emile Durkheim in the early twentieth century, and indeed since Buber, theorists have expressed the importance of connections between religion and social-emotional phenomena. A few theorists continue that trend today. But as I have indicated, to fully probe the origins of religion, we must look beyond even the first glimmers of *human* evolution to examine the emotional lives of the apes. And so I start the evolutionary clock earlier than do others who chart the origins of the religious imagination.

The challenge at the heart of this book is to tell the story of the *earliest* origins of religion. As is already clear, commitment to an evolutionary perspective on religion amounts to a claim that humans evolved God gradually and not via some spiritual big bang. Before moving, in subsequent chapters, to specifics of the evolutionary perspective itself, it remains to say something more concrete about religion itself. One linguistic clarification can be made immediately. By adopting the term "the human religious imagination," I do not mean to imply that humans simply make up God, gods, and spirits in their imaginations. Nor do I claim—nor, indeed, *could* I claim—that these sacred beings are real in our world. Matters of faith are not amenable to scientific analysis, experimentation, or testing; writing as a biological anthropologist, I remain agnostic on this question. My focus is on our prehistory, and on how—and why—we evolved God as that prehistory unfolded.

A BRIEF WORLD TOUR

Our exploration of the evolution of the religious imagination begins with visits to three locations across the globe. In West Africa's Ivory Coast, we walk through a lush rainforest and let our senses take us on a journey usually reserved for calling monkeys and swift-flying, colorful birds. Slowly we make our way under the thick, humid canopy, and find ourselves looking over the shoulder of a scientist who observes a female, not yet of adult age, laying motionless on the ground. She is dead, the victim of an attack by a leopard.[4]

In death, Tina becomes a magnet for other members of her community. Sitting around her body are twelve individuals, six males and six females. But these quiet observers are not people of the Senufo or Guro tribes, or indeed of any of the other human tribes in the region where Tina was born. They are chimpanzees, our closest living relatives in the animal kingdom.

Thanks to research spanning five decades by Jane Goodall and other scientists, the world has known for some time that chimpanzees are highly intelligent. They make and use tools in order to crack open hard-shelled nuts and to probe for hard-to-reach insects. Working cooperatively, they capture monkeys in shrieking hunts that end in frenzied sharing of meat. Deeply social creatures, chimpanzees express joy when they play together or reunite after long separations. Still, what happens next among the chimpanzees near Tina's body comes as a surprise.

Some of the chimpanzees stay with Tina's body for over six hours without interruption. None licks Tina's wounds, as these apes sometimes do when a companion is injured but still alive. Some of the males do drag Tina's body along the ground a short way, while other chimpanzees inspect, smell, or groom it. Brutus, the community's most powerful or "alpha" male, who had been a close associate of Tina's, remains at her side for five hours, with a break of only seven minutes. He chases away some chimpanzees who try to come near, allowing only a single infant to approach. This is Tarzan, Tina's five-year-old brother. Recently, Tina and Tarzan's mother died. Now, Tarzan grooms his dead sister and pulls gently on her hand quite a few times.

Brutus's behavior toward Tina's little brother indicates that he, Brutus, knew that Tina and Tarzan meant something special to each other. Taken together with other evidence to be reviewed in this book, including observations that I have made of captive apes in my own research over many years, this information suggests that Brutus was capable of feeling something like empathy. If so, Brutus was able to project himself into Tarzan's situation and imagine what Tarzan might experience at the sight of his sister's dead body.

Moving on now, we travel to Cameroon. Here and in neighboring Gabon, the Fang people clear land from the surrounding rainforest in order to grow crops. As we enter a village and observe the plantains and manioc under cultivation, we find that the area has more inhabitants than just the Bantu-speaking men, women, and children who live, work, love, laugh, and cry here. An array of spirits, both witches and ghost-ancestors—not all of them benevolent—lives here too.[5]

In Fang villages, witches may cause great woe and anxiety: crops fail and people die because of them. When witches congregate for banquets, the Fang say, they eat their victims and strategize about the atrocities they will commit in the future. Further, ghost-ancestors observe what the Fang do in their everyday lives, and have desires and make actions of their own. The world is affected directly by these forces. Sometimes, one's ancestors may intervene to help in a struggle against malicious spirits; at other times, living spirit specialists may intercede and, through ritual, work to help those at the mercy of witches. Still, the spirits are often beyond people's control, often quite frighteningly so.

The anthropologist Pascal Boyer draws a vivid contrast between the way the Fang view these spirits and the way they view their two major gods, Mebeghe and Nzame. Though Mebeghe created the earth and all its creatures, and Nzame invented houses, taught people to raise crops, and so on, "these gods do not seem to matter that much" in everyday life.[6] The spirits, by contrast, matter enormously. Rituals carried out by the Fang, indeed emotions felt by the Fang, center around these spirits.

An ocean and a continent away, in the United States, we make our last stop in Mississippi. Here, we eavesdrop as the tenant farmer Joseph Gaines talks with the psychologist Robert Coles about his faith: "I'll be praying to Jesus, and I'll feel Him right beside me. No, He's inside me, that's it. I think the church people, they want you to come visit them, and

This figure, a focus for prayer, guarded a box containing skulls and bones of Fang ancestors.
Werner Forman/Art Resource, NY

that way you meet the Lord, and His Boy, His Son. The trouble is, you leave, and the Lord and Jesus stay—they don't go with you. . . . So I say to myself: be on your own with God—He can be your friend all the time, not just Sunday morning.

"I'd like to be a minister, so I could know the Bible, and preach it to other folks. But in my heart, I don't believe the Lord wants me preaching on Sundays; He wants me living His way all the days of the week."[7]

Reading Gaines's words, two aspects of his faith become clear: it is *personal* and it is *active*. Though his love for the Lord and Jesus may be enhanced through Bible reading and churchgoing, its deepest source is the intimate relationship that he has with these beings. Further, the faith that this intimacy engenders is, and should be, expressed through action in the real world outside church walls. Belief and action are intimately entwined with each other and with the personal, and transform Gaines's life on a daily basis.

Each of these three vignettes offers a glimpse into a complex world of emotional relating. If one slips on the cloak of an anthropologist trained to think about broad patterns in human societies around the world, it's easy to see a link between the beliefs and rituals of the Fang in Cameroon, and the beliefs and actions of Joseph Gaines in the American South.

Certainly, the specifics of these two sets of beliefs differ. To many Westerners, the idea that modern people become seriously alarmed about witch banquets and ghost-ancestors may seem distinctly odd, even "primitive." To the surprise of no anthropologist, however, the Fang express a like incredulity when certain Christian beliefs are explained to them. They are "quite amazed when first told three persons *really* were one person while being three persons, or that all misfortune in this vale of tears stemmed from two ancestors eating exotic fruit in a garden."[8] In other words, the Christian ideas of God, Jesus, and the Holy Spirit as a trinity of beings, and of the Fall of Adam and Eve, seem distinctly odd to them.

Beneath the differences, though, at the level of meaning, there exists a fundamental similarity in the Fang and Mississippi examples. In both cases, people enter into a deeply felt relationship with beings whom they cannot see, but who are present daily in their lives and who transform these lives. I will have a great deal more to say about the character of this kind of relationship and its link to belongingness; for now, the key point is that an intimate social relationship between living people and supernatural beings of some sort is characteristic of human societies everywhere.

At first glance, however, little would seem to unite these two human-centered vignettes with the anecdote about the chimpanzees responding to Tina's death. Empathy is not religion, after all. Impressive though empathy may be as an indicator of emotion and intelligence—especially in nonhuman creatures—it is not the same as the capacity for faith or belief in something greater than oneself, nor as the capacity to reject this faith or belief. Chimpanzees do not spend years in studious contemplation of holy books; they do not enter into lengthy apprenticeships to learn how to become shamans and heal the sick in their communities; they don't build soaring, stained-glass-studded cathedrals, immense temples, elaborately tiled mosques; they don't debate agnosticism versus atheism. But, to revisit a point already made, I am confident that much in the behavior of Tina's community points us toward an understanding of what came later—religion. To move toward an understanding of why, let's examine now the ways in which the term "religion" is used when describing human beliefs and behaviors. First, we confront the most challenging question of all: What *is* religion?

RELIGION DEFINED?

"Operationalize!" As an anthropology graduate student at the University of Oklahoma, I heard this command ring out time and time again. My professors exhorted us, fledgling social scientists ready to initiate new research projects, to define our focus as precisely as possible *before we started*. This process, called operationalization, sounds simple enough, at least until you try it.

I learned this lesson in the mid-1980s, when preparing to travel from Oklahoma to Kenya to carry out my doctoral research. I had become fascinated with the question of how, in group-living monkeys, the younger generation learns about the feeding practices of its elders. This question loomed especially large for baboons, who live on the Africa savanna and select only the most nutritious parts of available grasses, flowers, and other assorted foods, while rejecting those that are potentially toxic. How do young baboons come to know which items and plant parts are safe to eat? I already knew that such abilities were highly unlikely to be instinctual, given the degree to which monkeys learn about other aspects of their world from each other. But how do they figure out how to prepare the food parts that they select, to ensure that they are as nutritious and safe as possible? This problem had been little studied in the field.

As I designed proposals to submit to national funding agencies, I discovered challenges to operationalization that were hidden within the most innocent-seeming concepts. How should "eating" be defined, for example? Did I want to claim that a monkey mother begins to eat only when food actually disappears into her mouth and chewing commences? If so, any behaviors involved in food preparation—peeling bark off a tree, or plucking spines off a fruit—would be excluded from my definition. Given that I wanted to understand how infants learned to prepare as well as to select food, wasn't such a narrow definition *too* narrow?

If, on the other hand, I began to clock eating from the moment a monkey touched the tree or fruit in question, what would I then do if she never actually ingested the item, but instead rejected it? Perhaps monkeys sometimes try to discover through touch as well as vision whether fruit is soft and ready to eat, or too hard and best left for later.

The more I agonized over these definitions, the faster the problems piled up. The Kenyan baboons I was about to spend my days with live in matrilines, social units organized around a core of female relatives. I was especially keen to find out whether infant baboons learned about feeding from their matriline partners: their mothers, grand-

mothers, aunts, and siblings. Or did they proceed on their own, learning by trial and error?

Social learning turned out to be even harder to operationalize than
eating. Consider an infant baboon sitting next to her mother. The baby
closely observes her mother eating the petals of a colorful flower, one
of many flowers that have sprung up, carpetlike, at the onset of the
Kenyan rainy season. Shortly thereafter, the baby eats petals from this
same type of flower. Does the infant's close scrutiny of her mother
count as social learning? What if, though, in toddling around near her
mother earlier that morning, the infant had sniffed and touched that
same flower all on her own? Would that prior and wholly independent
experience with the flower sway our judgment, leading us to suspect a
mix of independent and social learning? And how could I ever hope to
distinguish the two types of learning, given that my methods called for
me to observe one infant at a time for only fifteen minutes, and then
switch to observing the next?

Eventually, I worked out definitions that satisfied me (and, more
important, the funding agencies). In fact, these definitions worked ad-
mirably once I was actually in Kenya observing the monkeys.[9] But the
process of developing them was much more complicated and time-
consuming than I had anticipated.

And here is the point of my stroll down memory lane: my graduate-
school experience of defining baboon behaviors pales in comparison
with the challenges inherent in trying to define religion. Wishing to
operationalize what counts as religion in a single human community—
the Mormons of Salt Lake City, Utah, or aboriginal peoples living in
a certain region of Australia, let's say—is daunting enough. But I am
proposing to look at religion quite broadly; my project is to explicate the
origins of the religious imagination as expressed in near infinite variety
by human populations in the present and the past. It might be reasonable to wonder whether any single definition could possibly encompass
the religious beliefs and practices of Mormons, Australian Aborigines,
the West African Fang, Roman Catholics in the United States, and on
and on.

Anthropologists, naturally, are used to thinking through just this sort of challenge. They take care to avoid defining religion in ways that might be culture-bound or otherwise too restrictive. Most people in southern Virginia, where I live, would likely feel no unease at defining religion in terms of one's belief or faith in God.[10] And at first blush, this seems logical enough, especially if we make a minor amendment and add to "God" a phrase such as "or gods and other supernatural figures." This expansion would include the Fang's ghosts and witches, for example, and isn't religion all about belief in such beings?

But this amended definition does not apply in all cases; and, to have value, a definition must be more than a description that works well enough most of the time. A look at Buddhism reveals why religion cannot always be equated with faith in supernatural beings. The Buddha, or more correctly the man Siddhartha who became known as the Buddha during the course of his life, lived in India five hundred years before the time of Christ. Founder of a tradition that grew into the world's fourth largest religion, he believed neither in God nor gods. Today the vast majority of Buddhists also reject the notion of a powerful omniscient or omnipotent divine being. Instead, the focus in Buddhism relates to one's striving to overcome suffering and to reach enlightenment.

Just as the Bible is holy to Christians, the Torah to Jews, and the Koran to Muslims, so a sacred text guides Buddhists. Based on the Buddha's teachings, this text is called the Tripitaka, possibly because it was originally written on palm leaves stored in baskets called pitaka. Many of the world's traditions are "book religions" in this way. Yet it is easy to see that incorporating a sacred text into the *definition* of religion would be as unsatisfactory as incorporating a belief in God—even more unsatisfactory, given the multitude of religious traditions that flourish today in the absence of any writing at all.

One way to approach the problem of defining religion, then, is simply to go on listing the ways in which it's possible to go wrong by focusing too much on specific aspects, even those aspects familiar to Americans and Europeans. Boyer uses such an exercise to good effect.[11]

He conveys the diversity of beliefs and practices in the world's religions by enumerating eight findings of anthropology:

+ Supernatural agents are diverse, ranging from one unique god to many different gods, spirits or ancestors.
+ Some gods are believed to die, not to exist eternally.
+ Many spirits are "really stupid": that is, an all-knowing and all-powerful deity is not always the rule.
+ Salvation is not always a central preoccupation. Some people believe that the dead become ghosts or ancestors; little concern for the soul is in evidence.
+ Official religion is not the whole of religion. Various permutations of the official doctrines exist, so that even within religious denominations tremendous diversity in belief and practice can be found.
+ You can have religion without having "a" religion. Not everyone chooses to practice one among a smorgasbord of competing religions; often, embracing religion is not a behavioral choice by an individual but is the only way of being that's imaginable in a given community.
+ Similarly, you can have religion without having "religion." A linked set of beliefs and practices may be vital to a community without that community adopting a concept of "a religion" at work.
+ You can have religion without "faith." People may just simply know what they know (e.g., that ghosts exist among them) without approaching it as a matter of belief.

Boyer disaggregates what is (for most of us) familiar about religion from what is definitional. For those of us who grew up either accepting or rejecting an all-knowing, eternal God who helps us reach salvation, it may take extra work to think this way, but it's necessary to do so if we are to think productively about the origins of religion. It's worth noting that, despite the heterogeneity reflected in his list, Boyer accepts that we *can* illuminate the origins of the religious impulse. He even feels confident enough about his own theory to title his book *Religion Explained.*

Given their sensitivity to the heterogeneity and diversity of the world's traditions, how do anthropologists ever manage to define religion? Almost certainly, the most widely used and cited definition was proposed by the anthropologist Clifford Geertz: religion is "a system of symbols which acts to establish powerful, pervasive, and long-lasting moods and motivations in men by formulating conceptions of a general order of existence and clothing these conceptions with such an aura of factuality that the moods and motivations seem uniquely realistic."[12]

Geertz asserts here that the core of religion resides in symbolic representations. A hallmark of humanity is that we experience the world by representing the things most important to us by reference to other things. We may express love of country by honoring a flag, romantic love by pledging fidelity via a ring, love of God by sipping wine that is considered to be the blood of Jesus. In the case of religion, according to Geertz, humans make sense of the world through symbols.

Most fascinating, perhaps, is what Geertz omits from his definition. Nowhere does he mention faith, belief, God, gods, spirits, eternal life, souls, or sacred texts. Of course, anthropologists are a contentious lot. If one hundred anthropologists were assembled in a room, at least ninety would immediately wish to tweak, if not substantially alter, Geertz's definition. But few, I predict, would try to wrestle back into it the omitted aspects. A broad definition lets us cast the net wide and search for origins without unnecessary constraints; too narrow a definition hobbles this project from the start. Remember, too, that defining religion is only a starting point; from there we may discuss components that are variously *expressed* in religious behavior.

ON BELIEF, PRACTICE, AND THE SACRED

Deserving of special mention are two further aspects of the operationalization process. So far, I may seem to be endorsing a dichotomy of sorts. Do I mean that it is perfectly sensible to define the three major religions, and some others, in terms of belief in God, and that the sole reason to craft an expanded definition is to encompass other religions? Being good anthropologists, we do wish to include in a definition the beliefs and practices of the Buddhists (who believe in no God), the Fang (who believe in gods, witches, and ghosts) and other groups. But in fact, a dichotomy of this sort is too simple and ultimately misleading. This realization was brought home to me by reading the work of Karen Armstrong. *The Spiral Staircase*, Armstrong's intimate account of the pain and joy she experienced as she struggled with her faith, is aptly named for the halting twists and turns of her journey toward spiritual awareness.[13]

Because—fittingly, as we will see—Armstrong's insights derive from life experience, the maximum benefit is derived from her account by following along with the chronology she lays out. At the age of seventeen, Armstrong joined a convent. For the next seven years, she lived with a community of nuns, and lived, too, with anguish. Against all her young hopes and expectations of ecstatic transformation in encounters with God, she could neither pray nor experience God in any way: "I never had what seemed to be an encounter with anything supernatural, with a being that existed outside of myself. I never felt caught up in something greater, never felt personally transfigured by a presence that I encountered in the depths of my being."[14]

The miseries of a life promised to a God she could not feel became too much to bear; Armstrong left the convent. Rocky years followed, marked by periods of wrenching doubt, of outright rejection of God, and even of drawing markedly closer to religious ecstasy. This last occurred when she suffered a seizure, whose cause was later diagnosed as temporal epilepsy. At one point Armstrong decided flatly that she was finished with God. Yet, throughout these gyrations of the spirit, as she finished degrees at Oxford, undertook a fellowship in London, and began a career, her hunger for God persisted.

The career she chose—after a detour as a schoolteacher in south London—involved, as I see it, devotion to God no less than did her years in the sisterhood. This devotion was not now expressed through convent rituals and attempts at prayer, but through explaining her doubts to others and seeking to grasp the nature of other religious traditions. At the start, this "devotion" was fueled by skepticism, even cynicism and anger, about organized religion. Armstrong might quibble with my terming this path a career "choice." When her book about the convent years drew notice and the media sought her out for religion-centered projects, she was as surprised as anyone at how her life had turned out.

Early on, when preparing to fly to Israel to prepare a documentary about St. Paul, she encountered a point of view wholly strange to her. Another scholar remarked that in Judaism, theology is far from central. To put it more starkly, what individuals believed—or did not believe—about God mattered very little. What counted was not belief, but action or practice.

Armstrong recounts her reaction: "I stared at him. I could not imagine a religion without belief . . . my Christian life had been a continuous struggle to accept the official doctrines. . . . How could you live your faith unless you were convinced that God existed?"[15]

Time spent in Israel, where Judaism and Islam are as intensely present as is Christianity, pulled Armstrong further into a world of ideas about faith, belief, practice, and action. Yet her goals in making the St. Paul documentary remained as clear in her mind as they were specific: to expose Catholic theology as antithetical to the original teachings of Jesus, and to show that St. Paul, not Jesus, was the founder of Christianity. At this point, still equating religion with belief, Armstrong thought that she could shake up Catholicism itself if she could reveal mistakes in the received wisdom upon which its beliefs rested.

But this turned out to be just another twist along Armstrong's spiral staircase. Deciding to write a "history of God," she immersed herself deeply in understanding the sweep of Christianity, Judaism, and Islam. Now her goal was even starker than before: "God, of course, did not exist, but I would show that each generation of believers was driven to invent him anew."[16] What she learned, both before and after writing *A History of God*, shook her, resonated with what her colleague had told her about action and practice, and gradually transformed her even further.

Armstrong sums up her current understanding: "Religion is not about accepting twenty impossible propositions before breakfast, but about doing things that change you. It is a moral aesthetic, an ethical alchemy. If you behave in a certain way, you will be transformed."[17] At the heart of all major world religions, she notes, are empathy, and compassion expressed in daily connection with others. Religion *is* practice. Only recently, and only in certain parts of the world, has religion come to be equated with a series of cerebral propositions about God, propositions that must be accepted.

Gradually, then, Armstrong found her way back to a relationship with the sacred. At this point, committed to operationalizing terms as I am, I should spell out that my practice will be to shift among "religion," "spirituality," and "the sacred" as equivalent terms. Often, a distinction is made among these three, perhaps most vividly between religion and spirituality. As one perspective has it, "True spirituality comes from the inside out." The implication here is that spirituality is interior where religion is external, even at times imposed.

In modern America, religion is sometimes assumed to represent an opposite pole from genuine engagement with the spiritual or the sacred, because religion is equated with institutionalized, organized religion. "Religion," in other words, brings to mind the sex-abuse scandal among Roman Catholic priests or the crisis in Protestant denominations over gay clergy, as much as it does the experience of faith through prayer and compassion. An anthropological inquiry into the prehistory of religion, however, is a very different thing because it aims to search out the very first stirrings of any sort of spiritual imagination. Relatively indifferent to patterns in or development of modern institutionalized religions, it has far more to do with the experience of encountering the sacred in everyday life, as will become clearer as we go along.

But for the moment, back to Armstrong, who has traveled far from the time during her convent days when she considered herself a dismal failure because she could not enter into intimate relationship with God. All these years later, she experiences the sacred with a mix of openness and doubt, in a dance that shifts from understanding to mystery; it is no easier than ever to pin down the sacred, or to hold on to it. But now she accepts the mystery as inevitable and believes that the sacred surrounds her. "Our task," Armstrong concludes, "is to learn to see that sacred dimension in everything around us . . . we will catch only a fleeting glimpse of it—in the study of sacred writings, in other human beings, in liturgy, and in communion with the stranger."[18]

To sum up our definitional musings, then, religion is all about practice and emotional engagement with the sacred, as defined by one's social group; it is not *necessarily* about a set of beliefs concerning supernatural figures, though it may be that, too. Here we find useful parameters in our search for the origins and development of the religious imagination. It is possible, of course, to search through prehistory for evidence of belief in God or gods, other supernatural beings or forces and we will. But we needn't require finding those elements to judge successful an exploration into religious origins. *Most critical to uncover is activity, aimed beyond the daily here and now and toward a sacred dimension, that is expressed through social-emotional connection with other beings. We seek evidence that religion is an active expression of belongingness aimed at a spiritual realm.*

ON THE SACRED ALL AROUND US

A final consideration in our operationalization process is reflected in several entries on Boyer's list of eight anthropological findings about religion. The very attempt to define religion implies that religion is self-contained, separate from all other areas of life. Yet as I've already hinted, this is not an accurate picture.

I do not refer just to the habit of attending religious services or devoting oneself to sacred texts or good works on a certain day of the week: Friday, Saturday, Sunday or some other designated "holy" day. Some of us may invest more in religious activity on certain calendar days, but no one would claim that, as a result, we are necessarily aspiritual on a Monday. A concentrated flurry of religious activities during a certain day, week, or holiday period may be wholly consistent with a pervasive interweaving of faith into all other aspects of one's life. The words of the Mississippi farmer Joseph Gaines ("I'll feel Him right beside me. . . . He can be your friend all the time, not just Sunday morning") remind us powerfully of this.

I am talking about something more. In the United States, the query "What is your religion?" comes as no surprise. On the day we are born, our parents answer this question on our behalf, and persistent inquiry on the topic follows us as we enter school, begin employment, and seek health care and medical insurance. Rarely do we think twice about replying, even if the reply is "None." If we do hesitate or refuse, the reason is likely because we judge the question too invasive of our privacy, not because it startles or puzzles us. But anthropologists remind us that such a question would strike many people around the world as strange, even incomprehensible.

Comparatively few of the world's languages include a word that translates directly as "religion." Thus, a question that some people hear as perfectly sensible utterly fails to compute for many others. For the majority of the world's population, religion is simply continuous with the experience of living, day by day.

Native Americans say that the sacred is completely seamless with other aspects of their lives. George E. Tinker, a member of the Osage tribe, notes that in traditional Indian culture, "nearly every human act" is infused with attention to religious matters: "Most adherents to traditional American Indian ways characteristically deny that their people ever engaged in any religion at all. Rather, these spokespeople insist, their whole culture and social structure was and still is infused with a spirituality that cannot be separated from the rest of the community's life at any point."[19]

When Indian hunters killed a bison, they did so only after ceremonies and rituals designed to acknowledge the cost of the act to the animal. Even the harvest of cedar bark required, in some tribes, prayer and ceremony. But Tinker wants us to see beyond hunting and foraging, directly into the heart of daily life, in order to grasp the degree to which the sacred infuses even the most mundane acts.

The Osage, Tinker's own tribe, provide a good example. Among this group, the reality of the universe derives from both unity and duality. Two divisions exist, the Sky and the Earth, but their power comes only when they are brought together. Towns were split into "Sky" and "Earth" halves by an east–west road. Individuals in each half carried out their daily activities in different ways, so that in one, people slept on the right side and put on their right shoe first, whereas in the other half, the opposite practices prevailed. "As a result," Tinker writes, "even in sleep the two divisions performed a religious act that maintained their unity as duality as they lay facing each other across the road that divided the community."

As Tinker's words illustrate, Indians themselves use the seamless nature of the sacred with other realms of life to "mark off" their own worldview as distinct from that of the Anglo world, and of the non-native world generally. Another example comes from the "faithkeeper" for the Onondaga Nation, who remarked when addressing a gathering at the United Nations, "I do not see a delegation for the four-footed. I see no seat for the eagles."[20] It is perhaps hard to imagine the U.S. ambassador to the UN making this statement.

In fall 2004, the Smithsonian's National Museum of the American

Indian (NMAI) opened on the Mall in Washington, D.C. At that time, the Navajo ethnobotanist Donna House explained the principles of Indian cosmology that she used in helping to design NMAI's outdoor landscape. She wanted to create something sharply distinct from the regimented look elsewhere on the Mall ("tulips all in a line," in House's words). In fact, she wanted a habitat more than a landscape: an untamed place that reflects the dynamic and interconnected lives of its inhabitants, including butterflies, birds, and other creatures as well as plants.

"Plants," said House, "were here way before people. They know you, have a relationship with you. It's a sense of recognizing the plants, the animals, the insects, as beings."[21] I admit that when I started research for this book, I hadn't thought along the lines of human-botanical belongingness! That the relationship House describes involves the sacred rather than simply the intimate is clear when one explores the museum itself. A good number of the 800,000 items (not all on exhibit) in NMAI's collection are so imbued with sacred powers that some native people, rejecting typical museum-world parlance, refuse to call them objects. If an item contains an eagle feather, the spirit of the eagle itself may be present in the feather; if a shirt is made from antelope skin, a relationship to the animal from which it came remains embedded in the shirt. "There is a very important respect in which Native American people see objects as being living, as animate instead of inanimate," explained the NMAI's director, Richard West, a Southern Cheyenne.[22]

Caution is called for when trying to sum up something as complex as Native American religion. Especially in past centuries and decades, and sometimes even today, Native American lifeways are analyzed in frustratingly myopic or inappropriate ways, as if one unified set of actions and beliefs could possibly be said to characterize all tribes and individuals.[23] Indeed, my own phrase, "Native Americans say that the sacred is . . ." may be guilty of implying a monolithic view toward religion that does not exist. Still, guided by the words of Indians themselves, we can identify a reverence for all the world and its creatures as a common thread in Native American religion, while recognizing variable expression across Indian communities. At the heart of this proposed commonality is a forceful "felt connection" in which one cleaves with the world and does not hold oneself apart.

This sense of the sacred may generalize, at least in some ways, to other people around the world, those who do not participate in "book religions" but experience the sacred via close daily interaction with nature and the forces of the universe.[24] It is even tempting to draw contrasts between this nature-oriented view of the sacred and tenets of the three major religions. When the story of creation is recounted in the Christian Bible's Book of Genesis, for instance, the voice of God speaks clearly on the relationship between humans and animals: "Let Us make man in Our image, according to Our likeness; let them have dominion over the fish of the sea, over the birds of the air, and over the cattle, over all the earth and over every creeping thing that creeps on the earth."[25]

This notion of dominion might seem to specify a hierarchy of worth, control, and power, with humans at the pinnacle. Certainly it has been so used by millions over the centuries, to justify practices that subjugate animals to human will. Animal-rights activists point out, however, that one would be badly mistaken to read this passage as God's giving to humanity free license to control animals and make them do our bidding.[26] Noah's preservation of all animal species with his ark may be a singularly dramatic event, but many other Biblical passages too portray humans and animals, together, as God's creatures. We humans are exhorted to be merciful as God is merciful, and to love as God loves, which means mercy and love for all creatures of the Earth. In other words, "dominion" can be taken to mean compassionate care *for* animals, not control *over* animals.

Here we see that Biblical Christianity may diverge less radically from Native American cosmology (at least on this one issue) than is normally thought. At a more fundamental level, the drawing of hard and fast boundaries between the expression of one religion and that of another is always deserving of suspicion. Religion cannot be reduced either to a single way of believing or acting common to all adherents, or to sacred texts or authoritative pronouncements. The expression of

the religious imagination may not remain static even within a single individual, as we know from Karen Armstrong's memoir. Armstrong
climbed her spiral staircase, moving from strict doctrine within convent walls, to rejection of the sacred and cynicism about organized religion, to a joyful reunion with the sacred, experienced finally as
compassionate action.

While the notion of a life saturated with the sacred may be more
characteristic of certain approaches to religion than others, kernels of
this view are found everywhere. Great poetry conveys this sense elegantly, as in Emily Dickinson's "Some Keep the Sabbath Going to
Church":

> Some keep the Sabbath going to Church—
> I keep it, staying at Home—
> With a Bobolink for a Chorister—
> And an Orchard, for a Dome—
>
> Some keep the Sabbath in Surplice—
> I just wear my Wings—
> And instead of tolling the Bell, for Church,
> Our little Sexton—sings.
>
> God preaches, a noted Clergyman—
> And the sermon is never long,
> So instead of getting to Heaven, at last—
> I'm going, all along.

Dickinson has no need to survey the heavens in search of the sacred; it is all around her.

When, in "Song of Myself," Walt Whitman writes, "And the running blackberry would adorn the parlors of heaven . . . And a mouse is
miracle enough to stagger sextillions of infidels," his words span the
centuries and resonate with those of Donna House in speaking of the

Native American view of plants and animals. When we gaze at Vincent van Gogh's *Wheat Field with Crows* or his *Starry Night*, we may feel that humans are one small part of the universe, not perched at its pinnacle. When we visit Frank Lloyd Wright's Fallingwater in Pennsylvania, and walk among the trees, wildflowers, and boulders in order to admire the rushing waterfall, we grasp the meaning behind Wright's "I believe in God, only I spell it Nature." The visions of these artists give us a sense of continuity with beliefs and practices in other societies and perhaps in times past.[27]

Tracing the prehistory of the religious imagination is not equivalent, then, to explaining how churches, mosques, temples, and their associated sacred texts developed. Defining religion in a way that includes all human expressions of the sacred, past and present, differs from struggling to operationalize the behavior of feeding by African baboons. The very process pushes us up against not only a variety of languages, cultures, and customs, but also the need to capture the sense of wonder and awe at the core of what it means to be human in the first place.

In the end, we are left with two different things at once: a definition of what we mean (and do not mean) by the concept "religion," and a wish to allow for continuity between religion and other aspects of life. What must be kept in focus is humans' emotional relating to the sacred. The heart of the matter is located there, not in certain sets of beliefs about deities or supernatural beings.

How and why did humans develop an emotional connection to the sacred? In what ways did this connection play out in behaviors and practices related to belongingness, in the past? How did it develop from the emotional relationships among hominids and among the great apes, such as the chimpanzees who responded so strongly to their companion's death? These very questions take for granted that humans and apes did, indeed, evolve from a common ancestor. In the next chapter, I examine the relationship between apes and humans in more detail, and explore those ape behaviors that point us toward the earliest building blocks of the human religious imagination.

TWO

Imagining Apes

T THE EMOTIONAL CENTER of my life are animals, together of course with my (human) family and friends to whom I dedicated this book. With my husband and daughter, I rescue abandoned cats, turtles, birds, and the garden-loving groundhogs our neighbor vows to shoot. Daily, my moods are shaped by interactions with the four cats and one rabbit that share our home, and even by the events among the twenty-plus feral cats that we help to feed (and to neuter, so that no more homeless kittens will swell their numbers).

With one of our domestic cats I feel a special connection. After a troubling day, or just a tiring one, comfort arrives on my lap in the form of a jet-colored kitty named for a small, pretty town in New Mexico. Purring and kneading her paws into my legs, Pilar stares lovingly at me and I feel the tension in my muscles seep away.

My response to locking eyes with a gorilla or a bonobo is a little bit different. When spending time with the gorilla family I have filmed over the span of six years, I sometimes reflect upon an etymological nugget: the words *eye* and *window* are historically related. (English *win-*

dow derives from Old Norse *vindauga*, which translates as "wind's eye"). Shakespeare played on this link when he wrote, in *Love's Labour's Lost*, of eyes as windows of the heart.

In apes' eyes, I see something different from what I see in the eyes of any other animal. To be sure, when doing research in Kenya, I felt a sense of contentment at being so close to the monkeys, especially on days off when I could relax at a watering hole. The baboons, munching grass bulbs, flowed all around me, glancing at me and vocalizing softly to each other. I felt a connection with them, akin to but somehow *more than* the appreciation and wonder I felt as giraffes, warthogs, crested cranes, and even elephants also foraged around us.

What, is so special in the eyes of the gorillas at the National Zoological Park? Or in the eyes of the bonobos and chimpanzees I have studied elsewhere? Gazing into apes' eyes, I see reflections of emotional depth, a depth visible to anyone prepared to observe calmly and patiently. This is easiest to do at a good zoo and when the apes are not surrounded by noisy visitors who ignore ape etiquette. (If you're rusty on the fine points of ape etiquette, just follow the rules that your parents drummed into your brain when you were young: do not raise your voice to, throw objects at, or stare at those around you.)

What makes a "good" zoo is a complex question, but I have in mind zoos where the animals' needs count more than those of visitors. Apes do best in large enclosures that have not only access to the outdoors and many places to play and climb, but also bushes and grottoes in which to escape the prying eyes of gawkers. Should you be lucky enough to visit such a place, listen for soft, breathy laughter as juvenile gorillas chase and wrestle playfully. Look for the experienced chimpanzee mother who effortlessly coordinates her movement with that of her infant, so that when she flexes her knees and looks over her shoulder, the infant "reads" her posture and runs to climb upon her back. Watch as two bonobos negotiate the start of a grooming session, settling by head nods and shifts of posture and gaze who will be groomed first.

Although nothing replaces patient observation in person, a high-quality photograph may convey key aspects of an ape's emotional depth. Look, for example, at the eyes of the chimpanzee male Socko, so beautifully captured by Frans de Waal.[1] For a time, Socko was locked in rivalry with another male for his group's top-ranking or alpha position. During repeated conflicts, Socko experienced the force of his rival's formidable canine teeth.

Remarkably clear in his eyes is Socko's preoccupation, concern, even worry, over his struggle for dominance and its possible consequences. How precisely can a human so readily understand Socko's emotional state? The answer does not lie in some instinct encoded into either human genes or the human brain. Humans and the African apes grow up in groups where belongingness is central, so that keen attention is paid to social details. Humans and apes become able, through experience since earliest infancy, to read all the important nuances: the set of the brow, the tenseness around the eyes, the quality of the gaze itself. Humans, of course, add language to the mix, and surely exceed the apes in what they can read in the social partner. Still, gesture and

Worry is reflected in the chimpanzee Socko's eyes.   *Frans de Waal*

posture too can be eloquent, so that apes, too, intuit feelings in their social partners, feelings they may help to create or to shift by their own actions, whether excitable or soothing.

As de Waal has noted, when looking into a face such as Socko's, a human recognizes a fellow primate with a personality, a will, and an active mind. As it turns out, when Socko's social status improved, his worried look disappeared. Even though his tension was frozen in time by the camera's click, we can take comfort in knowing his mood did improve!

Less fortunate was a chimpanzee who lived at Gombe, Tanzania, a few decades ago. When, in the early 1960s, Jane Goodall first gained the Gombe chimpanzees' trust and could record from up close the apes' behavior, she became fascinated with a female she named Flo and with Flo's offspring. For years, "Flo groupies" around the world tracked reports of the family's activities. In 1972, when Flo died at the age of forty-nine, her obituary appeared in the *Times* of London.

But Flo's passing was soon eclipsed by a second death. Flo had spent her waning years in close company with Flint, her youngest son. Chimpanzee mothers are physically close with their infants for a prolonged period; indeed, the two are virtually tethered together as the baby first clings to the mom's belly and later rides like a jockey on her back. For about five years, the bond between the two remains the strongest of any in the group, although the infant's social circle gradually expands to include siblings, friends, and other adults. Flint's developmental trajectory, however, was different. At the age of eight, he was not yet fully weaned, an unprecedented situation (as far as we know) in chimpanzees. Dependent on his mother to an unhealthy extent far past his infancy, Flint simply could not cope when Flo died.

How heartbreaking is Goodall's description of Flint's decline: "He stopped eating, and he avoided other chimps, huddling in the vegetation close to where he'd last seen Flo. His eyes sank deep into the hollow sockets of his skull; his movements were like an old man's."[2] He became weaker and weaker. Three and a half weeks after his mother's death, Flint died too.

Through Socko's and Flint's eyes, we glimpse the emotional life of African apes. In this chapter we review the most compelling evidence for the deep primate roots of empathy, meaning-making, rule-following, and imagination, and how each may be an evolutionary building block of the human religious imagination. Are the African apes conscious beings? This question is either perfectly straightforward or entirely radical, depending on one's assumptions about human uniqueness and about the nature of consciousness itself. Discussing ape empathy is the first step on the road to answering this question.

EMPATHY

An article published forty years ago in the *American Journal of Psychiatry* gives insight into how empathy was once investigated in nonhuman species.[3] Scientists rigged up a system whereby two different amounts of food would be delivered to rhesus monkeys when they pulled two different chains. First, the monkeys were trained to pull the chains. Next came a test phase in which pulling the chain that resulted in the larger reward also caused a nearby monkey, visible to the chain puller, to receive an electrical shock. Once the monkeys experienced this contingency, two thirds chose to pull the chain that delivered the lesser (by half) food reward. Most striking was the behavior of two of the monkeys from the remaining third of the test group. Once they witnessed a neighbor monkey being shocked, they refused to pull either chain, one for five days and another for twelve days. Without a doubt, these monkeys suffered for their choice, as they did not get sufficient food to eat. In other words, the empathy they felt for their fellow test subjects overrode a basic survival need.

One might be tempted to conclude that the empathy shown by the rhesus monkeys outstripped that shown by the human experimenters, who devised and carried out the administering of electrical shocks to their fellow primates. Thankfully, primatologists now employ far less invasive methods for investigating the roots of empathy, both observationally and experimentally. And from their work we learn that as

striking as monkey empathy is, it does not match the empathy shown
by apes.

In mapping out the evolution of empathy, we could find no better
guide than Frans de Waal, a pioneer researcher in primate cognition.
Together with the psychologist Stephanie Preston, de Waal posits a
continuum of types of empathy.[4] All primates, indeed many mammals,
exhibit empathy that is based on emotional linkage with others of their
kind. *Emotional contagion*, for example, is a basic level of empathetic
response. When a monkey infant (or a puppy or a human toddler) expresses distress with her voice or body because she has become frightened or sustained a mild injury, other monkeys (or dogs or toddlers)
may become distressed, cry out, or show other forms of unease. The
"empathizer" in such cases may actually be unable to separate the distress of others from his own.

Learned adjustment may also occur. Imagine a young monkey, born
with a pronounced and permanent limp. When old enough to move
around on her own, she lags behind when her group flees a predator or
climbs a tree to sample some newly ripe fruit. After a few weeks observing this youngster's movement patterns, group members may begin to
adjust their pace of travel in some situations in order to let her catch
up, and may protect her in times of potential danger. In this case, the
shift is made on the basis of slowly accumulated experience, a kind of
trial-and-error adjusting of one's own behavior to that of another.

Apes go beyond emotional contagion or learned adjustment, Preston
and de Waal say, to a higher level, *cognitive empathy*. One ape can put
himself in another's shoes, intuiting the other's perspective or likely emotional state. For seasoned observers of apes, this claim is entirely reasonable.

One December morning in 2000, I was filming Kwame, the young
gorilla "star" of my gestural research, as he calmly swung from a rope all
by himself in the middle of an enclosure. His play, indeed the serenity
of the morning, was suddenly shattered by piercing screams. Kuja, the
silverback gorilla who was Kwame's father, had taken off after an ado-

lescent male, Baraka. Baraka ran away, screaming in fear. Though Kuja was the group's undisputed leader, Baraka had increasingly been challenging his position, typical behavior for a maturing male. The challenges that I had observed amounted to little more than adolescent harassment, such as throwing hay at Kuja and charging toward him. Zoo staff, however, had seen Baraka mating with the group's adult female, Kwame's mother—a far more serious infraction. Whatever had happened to set Kuja off on this winter morning, it went unwitnessed by me, but I did capture the ensuing fight on film.

Kuja and Baraka became locked in a behavioral dance, in which Baraka would show *some* of the signs of ape submission to a more powerful rival, but not all. The fear grin on Baraka's face, together with his arm-up defensive posture when Kuja came near, indicated that he had no intention of completely resisting Kuja's dominance. Neither, however, did Baraka "go prostrate" in complete submission: he never went lower than a sitting position. The conflict lasted longer than the fights in this gorilla family I have witnessed usually do, with Kuja eventually pulling at Baraka's lower limbs in what zoo staff and I interpreted as an attempt to force Baraka to "do the right thing" and literally lie low.[5]

Fascinating to watch as the conflict unfolded were the reactions of the other family members. Kwame, just over a year old, ran immediately to his mother, Mandara, and stayed glued to her, in a position of security and comfort. She, along with Kwame's older brother, Ktembe, ran right *into* the conflict. At first this may seem counterintuitive, and indeed not very clever, given that two comparatively large males were barreling around the cage, with bared teeth and aroused emotions! Shouldn't a prudent female or youngster get as far away from the action as possible?

These apes, though, are a family, and family members are emotionally connected to each other. When, during the height of the conflict, Ktembe, not yet four years old, intervened to swat at Kuja, and when Ktembe and Mandara together lined up behind Baraka, on the other side of Kuja (literally and metaphorically), the gorillas made visible this connectedness. More than emotional contagion or learned adjustment, the behavior of Mandara and Ktembe suggests that they could *empathize* with Baraka's distress as he faced the powerful group leader.

Can my interpretation be proven? Might I be overeager to ascribe complex abilities to favorite apes? Support for a cognitive-empathy reading of the National Zoo gorilla fight comes from Chicago's Brookfield Zoo. There, one summer day in 1996, a three-year-old boy climbed over a barrier and fell unconscious onto the concrete of the outdoor gorilla yard. Witnesses feared that the gorillas might harm such a vulnerable child. The eight-year-old gorilla Binti Jua surprised them; toting her own infant, she picked up the boy as well, then carried him to the enclosure's doorway so that zoo staff could easily reach him and take him to safety.

Binti Jua's actions were intelligent ones, not the results of "maternal instinct." The very notion of maternal instinct in apes has been discredited by case after case in which first-time mothers, lacking in social experience (as was allowed to happen in some zoos of the past) also lacked any child-rearing skills whatsoever. Like most gorilla mothers raised in social groups, Binti Jua was (and is) a good mother. Unlike most gorillas, she was catapulted into fame when videotape of the boy's rescue aired on television worldwide. Many people were astounded by Binti Jua's empathetic response to an injured child; ape researchers were not.

Instance after instance of cognitive empathy among both wild and captive apes could be described here. The famous "language" apes, those who use and comprehend human symbols, could certainly be included in such a catalog; any comprehensive account of chimpanzee Washoe, gorilla Koko, or bonobo Kanzi includes evidence of cognitive empathy. Washoe, for instance, once noticed an injury suffered by her closest human companion and caretaker. She was able to express her empathy using the elements of American Sign Language that she had mastered. The story goes like this:

Roger Fouts, at the time of this event, had for decades studied Washoe's linguistic development. But his relationship with Washoe, and with the other chimpanzees with whom she lives, extends far be-

yond one of researcher and subject. One Saturday, Fouts broke his arm
in a skiing accident. He arrived the following Monday at the chimpanzees' enclosure wearing a sling, and probably, he later speculated,
traces of pain on his face. "Instead of giving their usual, raucous, pant-
hoot morning greeting, they all sat very still and intently watched him.
Washoe signed HURT THERE, COME, and Roger approached and
knelt down by the group. Washoe gently put her fingers through the
wire separating them, and Roger moved closer. She touched him, then
kissed his arm. Tatu also signed HURT and touched him. What is
perhaps most amazing about their reaction was that ten-year-old
Loulis didn't ask Roger for his usual CHASE game until several weeks
later, when Roger's arm was on the mend."[6]

The weight of evidence for ape cognitive empathy, demonstrated
by events such as these, is considerable. Skeptics protest that each
event described is "just an anecdote." What is needed, they say, are multiple data points derived from hypothesis testing and statistical analysis on a single group. My response to this sort of methodological
complaint is offered at some length elsewhere.[7] In sum, the opportunity to respond empathetically doesn't come along all that often, at
least in the aftermath of rare events like conflicts or injuries. Nor is
there any guarantee that researchers will be observing when these
events do happen.

Further, some systematic research has been done. De Waal's studies on consolation offer convincing support not only for the presence of
cognitive empathy in apes, but for its near-total absence in monkeys.
Observing numerous aggressive incidents in a variety of primate
species, de Waal noticed a pattern: the victim often sought out the ag-
gressor, trying to make peace, rather than going off alone to recover and
avoid further trouble. This reconciliation behavior is interesting in its
own right. Even more fascinating, and directly relevant to empathy, is
that a *third* ape, one not involved in the aggressive incident, may console
the victim. Consolation may take the form of a hug, a touch on the
arm, a pat on the back, a bout of grooming or, as observed among
bonobos by Ellen Ingmanson, a direct gaze into the eyes.[8]

De Waal and his colleagues have gone beyond the reporting of single events to analyzing hundreds of postconflict periods in both monkeys and apes. With only a few exceptions, monkeys simply do not console each other after conflicts, apparently because they cannot project themselves into one another's emotional state. The chain-pulling monkeys described earlier probably operated on the basis of a form of emotional contagion activated when their companions were made to suffer an electrical shock. The difference between apes and monkeys is stark in this regard.

Empathy is expressed by apes in less dramatic circumstances too, as when they share knowledge with the next generation. In *The Information Continuum*, my first book, I theorized that the capacity for "social information donation" across the generations is much greater in apes than in monkeys. The baboon mothers I watched in Amboseli, Kenya, for instance, rarely altered their own behavior to intervene in their infants' food choices. Once, I watched an infant female baboon named Limau, just over three months of age, chew dried fronds and wood from the trunk of a palm tree. Her mother, only an arm's length distant (but facing away), did not intervene in any way as Limau did this, despite the fact that adult baboons do not eat these items. Three minutes after eating them, Limau vomited. Her mother did not respond to this, either. Though it is always possible that her mother simply did not *notice* any of these events—which in itself would be telling, given the tender age of her infant—my year's worth of data suggest that she is typical, for a baboon, in failing to intervene.

Ape mothers sometimes snatch away from their infants items that are edible but not part of the group diet. Even more strikingly, chimpanzee mothers facilitate the learning of tool-using and tool-making skills that can enhance ape foraging. At the Taï site in Ivory Coast, Christophe Boesch and colleagues have observed details of nut-cracking behavior.[9] Using stone or wooden hammers, chimpanzees aim precise blows at hard nuts in order to open them and extract the nutritious meat. Full mastery of the correct sequence of skills takes as long as eight years.

In season, mothers get a large percentage of their daily calories from nuts, and help their infants in age-appropriate ways that will be familiar to any parent or caretaker who has guided the long-term development of a skill in a child. Boesch's findings always remind me of the years my daughter and I have spent baking Christmas cookies together. When Sarah was very little, I'd do almost all the work myself. Her role was limited to dumping a cup of sugar (poured and measured by me) or a bag of chocolate chips (opened by me) into the batter (prepared by me)—while I verbally encouraged her and popped a stray chocolate bit into her mouth (and mine). A year or two later, I could leave a bag of sugar or chocolate chips, unopened, next to an empty measuring cup and empty mixing bowl, and let her do much of the work, although I still had to demonstrate, by example, the finer points of egg-cracking, so as not to end up with eggshell-laced cookies. Now, at age twelve, Sarah can carry out the entire sequence on her own, though she chooses not to work with the hottest of ovens. Soon she will take over that step as well.

At Taï, chimpanzee mothers intervene frequently in their offsprings' actions surrounding nut-cracking. In primatological jargon, they exhibit an average of twelve pedagogical interventions per hour of nut-cracking. Mothers routinely take into account, in other words, what their offspring can or cannot do at a particular stage of development. For the littlest ones, mothers simply hand over the occasional cracked nut, while allowing the offspring to observe their cracking technique. For older ones, mothers place the tools or the nuts near available anvils—and not just any tools or nuts, but those most likely to lead to successful cracking by the offspring. In two known cases, mothers slowly and carefully demonstrated how best to use a hammer after observing their offspring having difficulties. Modifying their own behaviors in these varied ways, the mothers show a form of cognitive empathy, in this case not about upset or injury but about what is known or not known.

Lest monkey partisans become indignant at my relentless separatism, I should emphasize that *some* monkeys in *some* situations do perform cognitively at levels comparable to apes. Monkeys occasionally do console victims of aggression, or teach their infants survival skills, or take the perspective of another monkey. What I'm after, though, is the behavioral pattern, and here we're on certain ground in insisting upon a distinction between monkey and ape empathy.

In *The Information Continuum*, I envisioned the evolution of teaching and perspective-taking from monkey to ape to hominid to human as a relatively gradual, step-by-step process. Incremental changes over time were caused, I hypothesized, by shifting selection pressures (themselves owing to environmental change) coupled with increased brain size. Another possibility exists, however. When change is rapid and unpredictable rather than gradual and incremental, behaviors may be the product of *emergence*. Preston and de Waal say that cognitive empathy emerges with other sophisticated "markers of mind" such as the ability to recognize the self in a mirror and the ability to deceive, and is "greatly increased" in humans and apes. I will return to the topic of emergence later. For now it is clear that, whether cognitive empathy evolved gradually or more rapidly, it represents a turning point in the evolution of social-emotional relating between social partners.

De Waal, in fact, considers nonhuman primate empathy to be a building block of morality: "I hesitate to call the members of any species other than our own 'moral beings,' yet I also believe that many of the sentiments and cognitive abilities underlying human morality antedate the appearance of our species on this planet."[10] Here is a link between primate empathy and the origins of religion. Acting toward another based on a "reading" of how that social partner might feel is an enormous evolutionary leap. While accepting the term *cognitive empathy*, I stress that what's really involved is cognitively channeled emotion, which may be expressed as caring or compassionate acts in the context of belongingness.

Many religions have a version of the Golden Rule: "Do unto others as you would have others do unto you." A moral injunction such as this underwrites the compassionate practice at the heart of many religions, and is easier to fully grasp when you can project your own inte-

rior world onto that of another person. Cognitive empathy has deep evolutionary roots; at some point during human evolution it combined, I believe, with other foundational elements to push human ancestors toward the expression of a religious imagination. Let's continue to survey the roots of these elements in apes.

MEANING-MAKING

Imagine two young chimpanzee males in Africa, reuniting after spending most of a day in separate traveling parties. The chimpanzee community to which the two belong is going through a slightly unstable period in which two other males are vying for the alpha or top rank. The two reuniting males (let's call them A and B) have been caught up in the instability that prevails as each rival seeks out allies to back up his own position.

Approaching each other now, A and B both appear tense to a seasoned observer; their hair stands slightly on end, their limb muscles are fairly rigid, and their gaze darts nervously around. Grunting softly, A slowly extends an arm, palm up, toward B. As (not after) he does this, B moves toward A, at first hesitantly and then more confidently. The two embrace each other and begin mutual grooming.

Typically, a primatologist studying communication would focus here on chimpanzee A as the *sender* of signals and on B as their *receiver*. In other words, an individual ape conventionally is thought to produce a signal that transmits meaning across space and time (however short either may be) to the communicational partner. That meaning may be transmitted by individual A and then extracted from the signal by individual B is an assumption that permeates much theorizing about human communication as well. This "transmission metaphor" greatly oversimplifies what language is all about; does it work any better for ape communication?

Remember that, in the sender-receiver approach, the grunt and the arm extension themselves are seen as conveying a message from chimpanzee A to chimpanzee B. Perhaps the message would be translated

as, "Come closer and all will be well between us," or, at least, "Come closer, I won't attack." In any case, it resides *in* the signals, just waiting to be extracted and comprehended by the receiver. Certainly, the body posture of the two chimpanzees might be taken into account by observers, and perhaps even the recent social history of the pair or the events going on around them. But these features would be relegated to a secondary role as external variables, with the focus put squarely on the linear act of ape-to-ape signal transmission.

But the sender-receiver approach fails rather spectacularly to do justice to the intensely social basis of chimpanzees' intelligence and emotionality. It is here that my own primate research enters the picture. Over the last decade, I have observed and filmed captive gorilla and bonobo families in order to investigate a new hypothesis: co-regulated communication in African apes may result in creative meaning-making. I believe that apes communicate in far more complex ways than the sender-receiver approach can capture. In this section, I "unpack" what the meaning-making hypothesis is all about. Taking to heart Chapter 1's lesson about operationalization, I start by defining what I mean by co-regulation.

My mental image of co-regulation is captured by a phrase that I turned into a book title: *The Dynamic Dance.* When apes or humans communicate, they adjust to each other's actions and choices moment by moment, just as one dancer subtly shifts the placement of a hand, or the speed and angle of a turn, as her dance partner shifts his. Co-regulation, then, is the unpredictable and contingent mutual adjustment between partners.

Among humans, even speech is co-regulated. The linguist Charles Goodwin has shown that speakers alter their sentences *as* their listeners begin to respond. Even when we are not aware of it, this process takes place routinely as we communicate. Let's say that I meet up unexpectedly with a friend while doing the week's grocery shopping, and launch into an effusive recital of all the recent good news in my life. Midway through my first sentence, though, I notice that my friend is glancing away toward the vegetable bins. She will not meet my eyes,

and slowly turns her body half away from mine. Without planning to, I begin to edit myself: I end my sentence prematurely and start a new one, forming it into a question about my friend's health.

Compare this outcome to an equally hypothetical one from the day before. I had begun then, too, with a similar recital of my good news. In that case, my conversational partner, a colleague at work, looked directly at me and smiled, slightly lifted her eyebrows, inclined her head in my direction, and even moved her body slightly closer to mine. Confident in a continuing warm reception, I proceeded along the same conversational track for two more sentences. In sum, a single starting point had two divergent outcomes because in each case I responded contingently to my social partner. Both conversations were co-regulated.

Human communication is about far more than speech. This, too, is clear not only from my examples, but in everyday life, perhaps never so vividly as when we interact with an infant. It is no coincidence that the term "co-regulation" was coined by Alan Fogel, who studies infants. Between infants and their caretakers, as Fogel shows, we can see an elemental emotional connection and mutual adjustment of action at a very basic level.[11]

When she was a tiny baby, my daughter Sarah's favorite activity was the "lean game." Though we invented our own version, no doubt families in many societies engage in a variant of this game. As Sarah sat on my lap, I inclined my upper body and face toward her, forming an exaggeratedly delighted face (open mouth, bright eyes, uplifted eyebrows), and making happy noises. Precisely as I began to lean, I gently pulled her upper body and face in toward me—and watched her form an equally delighted face and make equally happy noises. This could go on for quite a few rounds, with our emotional arousal building to a high pitch. When she was a little bit older, Sarah could propel herself toward me under her own power, and adjust her own actions to mine more precisely.

The co-regulation in this example may be unusually explicit, but the underlying principle of co-regulation is the same across all types of human communication. Our posture, our gesture, our facial expression, and the direction and intensity of our gaze all affect our partner moment by moment, and not just in a linear way. Co-regulation is not a stimulus-response chain of events. It's not "He looked away and then I softened my voice tone," but rather "As he began to shift his gaze away, I dropped my voice and spoke more softly." It's not "She turned to me, and then I smiled," but rather, "As she began to turn her torso in my direction, I started to smile." A web of contingencies, overlapping in time, characterizes co-regulated communication.

Humans make meaning together when we communicate: sometimes more fluidly, sometimes more haltingly; sometimes smoothly and sometimes only in fits and starts and with the need for repair. Descriptions of how we make meaning tend to be ignored by theoretical linguists, who focus on sentence structure, recursive syntax, and the conduit model with its senders, receivers, and signals.[12] Yet meaning-making lies at the very heart of human social relating, a point captured by Alan Fogel when he stresses that "the creation of meaning is the motivation for communication and for the persistence of relationships over time, not the mere meeting of needs through other people."[13]

My aim is to show that a basic kind of meaning-making can be found in African ape communication, a fact that can be seen by taking as the unit of analysis not the signal but instead the social event. Typically, primatologists take behavior like an arm extension, a head nod, or a facial expression as the thing to look for, to count up, to analyze statistically. With this kind of laser-beam focus on the individual components of some interaction, it's easy to miss the fact that only when social partners come together and play out an interaction does the arm extension, head nod, or facial expression take on communicative sense. Over the long term, social events build up between social partners to create social relationships, and it is here that we can reconnect to long-term developing emotional patterns.

Tellingly, the terms of my own writing have shifted over the years: from an emphasis on the transmission of information by an exchange of signals from one animal to the next (*The Information Continuum,*

1994) to an emphasis on dynamic, always-shifting mutual adjustment between partners (*The Dynamic Dance*, 2004). Fogel's research on
mothers and babies has remained a touchstone for me as I negotiate
this new path. His work reminds us that the family is a system.[14] When
one family member behaves or communicates a certain way, the entire
family is affected; a change in one person's behavior or communication
changes what goes on in the family as a whole. This model has powerfully affected the study of healthy and dysfunctional families and of developmentally normal and developmentally challenged children.[15] In all
cases, systems thinking leads us to see that meaning in our everyday
lives emerges through the human need for belongingness and its expression via emotional connection as we interact with others we care
about.

All this would be an interesting enough academic exercise, but a
sterile one for the evolutionary perspective, if it failed to match what
apes actually do in the real world. Ape-watching has taught me that,
as I have hinted, the concept of co-regulation beautifully describes
what happens in the lives of apes, and further, that this co-regulated
meaning-making starts very early in life.

I inaugurated my ape-communication research at the Language
Research Center near Atlanta, then home to Kanzi, the famous
symbol-using bonobo. Kanzi had been raised by ultra-competent "supermom" Matata, who had adopted him in infancy. Years later when I
arrived at the LRC, Matata was pregnant for the fourth time. My student Erin Selner was lucky enough to film the birth of the new bonobo
baby, and in the ensuing weeks, we documented the earliest mother-
infant interactions in this highly intelligent species.

A favorite event of all my ape-watching years happened when the
new bonobo infant, Elikya, was just nine weeks old. Before she was able
to walk or even crawl steadily, Elikya was able to create meaning together with Matata. Here is what our video record shows:

Elikya sits indoors, near Matata and Elikya's five-year-old sister
Neema. Matata takes Elikya and hands her over to Neema. As she is
transferred, Elikya makes a facial pout in her mother's direction.

Neema holds her, but Elikya extends her arm three times in succession back toward Matata. Although Elikya is close enough to touch her mother, she makes these movements instead, quite slowly and deliberately. As Elikya is in the process of making her third arm extension, Matata takes her back. As Elikya relaxes against her mother's body, Neema pats her gently.

When caught in a situation distressing to her, Elikya learns—at a very tender age—that she can help to bring about, through the movements of her limbs and face, a change in a social event. Elikya's specific intentions when she made a pout and extended her arm cannot be known to a certainty. It is abundantly clear, though, that together, mother and daughter responded to each other's contingencies. They converged on a meaning for Elikya's arm-extension: "Take me back."

Over the ensuing months, Elikya gradually built upon this success to coordinate increasingly well her actions with those of her older siblings and mother.[16] When she interacted with Kanzi, I was especially interested to see how well-coordinated the pair was; Kanzi is a bilingual ape, in the sense that he has mastered co-regulated communication not only with other bonobos, but with humans as well. Raised in a highly enriched environment, he is able to understand spoken English to a degree that astonishes many people. Kanzi responds to his human conversational partners via a wide range of vocalizations and gestures and by pointing to computer-based symbols called lexigrams.[17]

The dynamic dance of ape communication, then, develops from the emotional and embodied connection between ape infant and primary social partners. I have been able to document this process in greater detail among a second family group of apes, western lowland gorillas living at the National Zoological Park, part of the Smithsonian Institution in Washington, D.C. I began filming Kwame—the youngster who ran to his mother during the conflict between gorillas Kuja and Baraka, recounted above—on his fourth-week birthday in December 1999.

Like Elikya, Kwame is fortunate to have been born to a "supermom," a female with much success in raising offspring. Mandara is, in

A family of western lowland gorillas at the National Zoological Park, Washington, D.C. Left to right: Kwame, Kuja, Mandara, Kojo.   *Cindy Baker, College of William and Mary*

my eyes, the perfect primate mother, calm and loving. Her infants feel free to wander away from her and to explore the wonders of their physical and social environment, scampering back to Mom if they hear a loud noise or just manage to spook themselves by going too far from her. Eagerly, they investigate everything in their world: how it feels to roll down a grassy slope outdoors; what happens when you run and slap the glass window indoors, and how zoo visitors sound and act when you do this directly in front of them; how close you may run, jump, and play near Dad, the big powerful silverback leader of the gorilla group, without eliciting a loud threat bark or brief scary lunge.

Combined with this laissez-faire attitude is Mandara's laser-beam focus on her infants' welfare. She seems to judge precisely when and how to intervene in her infants' minor misadventures, and can now allow events to unfold to a specific point, one that enables the infant to gain valuable life experience but that does not threaten his physical safety or emotional security.

Once, Kwame's older brother Ktembe snatched him up and tried to make off across the cage with him. Though older, Ktembe wasn't all that much bigger than his seized charge, and certainly he wasn't an experienced baby carrier! As the two moved awkwardly across the cage, I could hear both Kwame's shrieks and the *thwack, thwack* of his body, dangling from Ktembe's arms, smacking into various objects, including a hard ledge. Mandara calmly moved up close to her two sons and stood still; Kwame broke free (perhaps his mother's presence caused Ktembe to loosen his grip) and ran to her. In this way, Mandara allowed her older son to gain some experience without allowing her younger son to experience too much distress.

Once tiny and uncoordinated, flinging his limbs around as he struggled to pull himself up his mother's body to nurse, Kwame has matured into a juvenile who struts his stuff in front of the "big boys" in his family. One of my research goals has been to describe the development of Kwame's requesting behavior. How does this young ape learn to ask for what he wants rather than just trying to grab it? Does the trajectory of his requesting behavior mirror that for his young brother, Kojo? To understand what I'm after here, think of a toddler who spies a cookie in his dad's hand and reaches out to snatch it. Then think of an older child who hungrily eyes the cookie, but extends his arm and says "Cookie?" rather than making a grab for it. How does a nonverbal counterpart to this developmental transition take place in apes?

The answer for apes mirrors the one for humans: requesting develops from an emotional and embodied basis with the mother (or any primary caregiver). Over the years, Kwame has regularly made the type of arm extensions to Mandara and to his other group members that so struck me when Elikya, the bonobo infant, was nine weeks old. Yet gorillas are not bonobos, and Kwame was not as precocious an infant as Elikya. His requesting behavior developed quite gradually, embedded in communication patterns typical of his species.

In the first months, Kwame clung tenaciously to his mother's body. At this time, co-regulation between mom and baby amounted to Kwame's shifting his body and limbs around on his mother's body, without benefit of much motor control. Mandara adjusted her actions to his: co-regulation was asymmetric, led by mom, but a dynamic dance nonetheless. Gradually, like any growing primate, Kwame began to toddle away from the safety of his mother. As he explored the wider gorilla world, new opportunities came along for co-regulated communication.

The notes I made while watching a social event unfold when Kwame was nearly six months old capture this change. Kwame sits near his mother on the cage floor. Using Mandara's body as a bridge, he climbs up the cage mesh. He puts one foot down on his mother's shoulder. Mandara opens her mouth (a mild threat), but Kwame is facing away and cannot see this. After some other contact between the two while Kwame is hanging on the mesh, Mandara shuffles a short distance away. As she moves, Kwame extends his hand, palm up, toward her, but she cannot see this. Kwame moves closer to her, still on the mesh, and touches her back several times. He could easily climb on her body, but does not. Over the next few moments, Kwame reaches toward and touches Mandara some more, and stretches his body toward her. She looks over her shoulder at him, and raises one arm up a bit, and turns. As she does this, Kwame makes a distressed-face expression. (It is not clear whether this expression is visible to Mandara.) As Mandara completes her turn toward him, Kwame makes what seems to be a pout face, and reaches for his mother as she brings him in to her body.

Once again, Mandara encouraged Kwame's independence and problem-solving skills, but intervened before he got into real trouble or became seriously upset. As the months went by, Kwame began to make arm extensions in the air—that is, in the space *between* himself and another gorilla rather than *on* his partner's body. These included the arm-out, palm-up requests so familiar to observers of adult apes. What fascinates me is not the simple fact that Kwame came to make this gesture, which is just typical ape behavior, but that it was truly a co-regulated process that unfolded with his social partners gradually over time.

I described the nuances of Kwame's requesting behavior in *The Dynamic Dance*. What's significant for our purposes is that Kwame came gradually to negotiate outcomes with his family members, and thus to make meaning with them. Even in this one zoo-living gorilla family we see much more than an exchange of signals by a sender and receiver!

Primatologists who study chimpanzees, bonobos, and gorillas in Africa do not use terms like "co-regulation" and "meaning-making": most were trained to use, and still accept, a traditional sender-receiver framework in which the meaning is thought to be contained in signals. But my scrutiny of the data convinces me that these phenomena exist in the wild just as in captivity—not for every instance of communication, but often enough.

Unsurprisingly, given what we know about the abilities of the young captive bonobo Elikya, wild bonobos make ideal examples. When two bonobos communicate in the forest, the meaning of their interaction is derived from a complex constellation of factors: the degree and intensity of the gestures both make, the identity of the participants, the participants' recent social history, what else is going on in the group, and, if one or both of the participants are male, whether he or they have an erect penis. (Famously, bonobos have a rather sexualized approach to the world, within as well as between the sexes.)

This perspective comes from the work of the primatologist Suehisa Kuroda, who has carried out field studies in Africa on the bonobos' rocking gesture, in which the apes rock the upper halves of the body and their head, sometimes while making arm raises or other gestures. Depending on the particular constellation of the factors described above, the meaning of the rocking shifts, and the behavioral outcome varies from sex to grooming to aggression. As Kuroda says, "The meaning of the interaction changes depending on the group situation and intensity of rocking."[18]

Judging from field reports, co-regulated communication and meaning-making happen regularly among wild African apes. And just as with cognitive empathy, indeed in part because of what we know about cognitive empathy, I believe that meaning-making among apes exceeds that among monkeys. The logic is this: in apes, cognitive empathy goes beyond mere learned adjustment, in that it is based on perspective-taking. This empathy then informs the apes' co-regulated communication, so that social partners go beyond moment-by-moment adjusting to a deeper emotional co-creation of meaning.

Let's step back now and look at the bigger picture. Fundamentally, the quest for the sacred is about a search for meaning in the deepest sense, going beyond making sense of shared communication between social partners to encompass a search for making things *matter*. Often, this idea is expressed as a process of *seeking in order to find*. It may be said, for instance, that humans seek God, gods, or spirits in order to find this deep level of meaning in their lives. But it may be more apt to say that humans seek God, gods, or spirits in order to co-create meaning with sacred beings.

How many of us have failed to wonder about the purpose of our lives? Our common humanity translates to a common desire to raise our children with love and care, to contribute to society through important work, to give back to the world through compassionate action, or all three. And woven throughout these actions is the need for belongingness, which may go beyond the daily routine with family, friends, and co-workers and extend to a transcendent level. That the "social partner" in this case is invisible, and ineffable, brings to co-regulation a different quality, just as meaning-making is brought to a deeper level.

Prayers may be answered or not; the performance of a sacred ritual may succeed or fail to bring about a desired result. But at bedrock is the belief that one may be seen, heard, protected, harmed, loved, frightened, or soothed by interaction with God, gods, or spirits. True, indifferent gods do exist, but they pay a price for their indifference. Many religions have a Sky God, a kind of ultimate ruler and arbiter of all things, who may be appealed to in times of crisis. Yet beings of this nature remain apart from people's daily lives: however powerful, they simply do not touch people at their emotional core. Significantly, Sky Gods tend over time to become fairly marginal and powerless figures, no matter how primary their role in the creation of the world. As

Karen Armstrong puts it, if a mythic representation of the sacred "does not enable people to participate in the sacred in some way, it becomes remote and fades from [people's] consciousness."[19]

People crave active participation with the sacred, a goal that may be achieved through participation in ritual. "The meanings of ritual are not private messages between two individuals," writes the anthropologist Richley Crapo. "Rather, these meanings embody important information about the ideological and social unity of the group; participation in rituals is an act that reaffirms this unity to those who already share the meanings."[20] Ritual helps to create order and certainty while tapping into deep emotions.

Unsurprisingly, many anthropologists consider the deep meaning-making of ritual to be uniquely human. People, and only people, create abstract religous symbols.[21] Certainly, ape meaning-making does not extend to creating such symbols or to deriving spiritual meaning from participation in ritual. Nonetheless, the data we have so far reviewed from our closest living relatives show that humans are evolutionarily primed to make meaning with our social partners through co-regulated communication. When combined with cognitive empathy, and with other tendencies I will now discuss, this basic meaning-making becomes a solid platform for the evolution of religious imagination and its own brand of very powerful meaning-making.

FOLLOWING THE RULES

Some years ago, an undergraduate student whom I knew only slightly began to address me, when we met in the halls, as "Barbara." This startled me, because at William and Mary, students routinely use their teachers' formal titles ("Doctor" or "Professor"). When I urge students to use my first name, either because we are working on research closely together or because they have graduated but stay in touch, it often involves a real effort for them at first. And here was a student I hardly knew, going out of his way to use my first name, uninvited and with apparent confidence!

What was going on? The student had violated a rule. I noticed this, and felt slightly uneasy about it; I am certain that uneasiness communicated itself to the student through my posture and facial expression. His violation was minor, to be sure. He had neither violated the College's honor code nor confessed to some criminal act. But he did break a social convention—at least, one that existed in the small universe of a traditional college in southern Virginia. And in so doing, he violated my expectations, and those of the college community.

Only later did I discover that this rule-breaker was enthusiastically fulfilling a class assignment. A fellow professor had assigned to her introductory students a fascinating task: they were to identify a social rule at work; flout it; observe the results; and report back to class. I never learned what grade the student earned for his "field work" of unnerving the faculty, but I do know that my colleague conveyed to her students something essential about human nature. We pay close attention to the conventions we collectively create.

Trained to think in evolutionary perspective, I wondered: Do apes follow rules? If so, to what degree? *Can* rules be understood in the absence of language? Why not, if some basic level of meaning-making occurs in the absence of language?

We have already glimpsed the emotional connectedness expressed in a gorilla family during a conflict that arose between the group's two oldest males. The silverback Kuja chased and threatened the subadult Baraka, who offered a mixed response: while acknowledging Kuja's dominance in some ways, he refused to submit totally. Kuja tried repeatedly to force or pull Baraka into a lower, more submissive posture. Given what is known about African ape cognition, I suspect Kuja's actions reflected his expectation about what subadults *should do.* Baraka, in other words, broke a social rule when he resisted Kuja, the group's leader. Though other group members supported Baraka and even tried to intervene in Kuja's actions, their effect was minimal—unsurprisingly, given the differential between their size and Kuja's.

When females and males are closer in size, and when the group is organized a bit differently, conflicts may take an interesting turn. Recall Socko, the chimpanzee photographed by de Waal, whose worry showed in his eyes. One day, as an adolescent, Socko mated with a female highly preferred by the group's top-ranking male. No slouch in the intelligence department, Socko (and the female) had met out of sight of that male, Jimoh. Unfortunately for the furtive pair, however, Jimoh was smart, too, and he searched the two out. Jimoh chased Socko aggressively; Socko ran, screamed, and defecated in fear (as Baraka had done when pursued by Kuja).

What happened next sets this chimpanzee event apart from the gorilla fight, however. Several females in Socko's group began to utter a bark that, de Waal reports, is used against aggressors and intruders. "When others joined in, particularly the top-ranking female, the intensity of their calls quickly increased until literally everyone's voice was part of a deafening chorus. . . . Once the protest had swelled to a chorus, Jimoh broke off his attack with a nervous grin on his face: he got the message. . . . These are the sorts of moments when we human observers feel most profoundly that there is some moral order upheld by the community."[22] De Waal believes that the chimpanzees were observing a prescriptive rule: Jimoh was going too far in asserting his dominance. Limits exist, and the group let him know it.

More than a mere pattern—that is, more than a behavior that most indivuals in a group do most of the time—a prescriptive rule is *enforced* in some way when violated. Just as with cognitive empathy and meaning-making, rule-following can be expressed in degrees. Various methods of enforcement exist, too; not all are as physical and aggressive as the actions carried out by the chimpanzee Jimoh and the gorilla Kuja.

In asking whether bonobos might follow social rules, Stanley Greenspan and Stuart Shanker sum up what full expression of rule-following would amount to: "We need evidence of instruction directed to young or new members of the group about the nature of the rule and when and where it applies, as well as evidence of sanctions against those who break the rule, and perhaps even awareness on the part of the rule breaker that a rule has been broken."[23]

Reviewing social events among bonobos in the wild as well as in captivity, Greenspan and Shanker conclude that "intriguing signs" of ape rule-following do exist. They report a case study among bonobos at Wamba in the Democratic Republic of the Congo. When a juvenile took a piece of sugar cane before the alpha male signaled that subordinates might eat, he was mildly disciplined by the alpha. A recent study indicates that some chimpanzee youngsters adhere to social rules when playing near to adults who might monitor their roughhousing or even punish them if things got too rowdy.[24] In the next few years, rule-following should open up within primatology as a legitimate area of study.

What, though, about symbolic rituals? Do any monkeys or apes participate in some form of these? Anthropologists John Watanabe and Barbara Smuts make a provocative claim along these lines for Kenyan baboons. When two male baboons greet each other, they use specialized communicative gestures and stylized body language. Members of a pair approach each other with an unusually stiff-legged gait, and one male may smack his lips, flatten his ears, and narrow his eyes. If his partner responds in kind, the two may then hug. Often, one male mounts the other, and a bout of mild penis-pulling may ensue before the pair splits up.

I am keenly interested in this report. First off, it's clear how much these baboons trust each other, because greeting males put their reproductive futures into their partners' hands in a very direct way. Most significant for an exploration of the roots of religion is Watanabe and Smuts's insight that the greetings are symbolic in nature. Elements of the ritual are borrowed from their original context and used in a new way. The lip-smacking, ear-flattening, and eye-narrowing sequence is borrowed from nurturing communication between a mother and her infant, and the mounting is imported from male-female mating behavior.

In a fascinating paper, Watanabe and Smuts explore some of the social implications of these greetings.[25] For us, a take-home message of their work is that there clearly exists some degree of continuity in symbolic ritual among nonhuman primates and among humans. And, of course, the baboon example situates that continuity in an emotional context: greetings *matter* to these males.

That religion is built fundamentally upon belongingness is a thesis of this book. As already noted, social ritual is a key to the expression of the religious imagination. And ritual in humans cannot be performed any old way, according to individual whim or desire; it unfolds according to precise rules devised by the community. This holds true even when an individual is utterly alone, in church or at home, and performs some component of a ritual.

Muslims may gather together to break the Ramadan fast, and a devout Catholic may recite the rosary all on her own; taking an evolutionary perspective allows us to imagine a link between those behaviors and what our closest living relatives do.

IMAGINATION

Robust evidence exists for cognitive empathy in African great apes, and new research points to some form of meaning-making and rule-following in chimpanzees, bonobos, and gorillas. Some primates even engage in symbolic rituals. All these behaviors can be interpreted, as indeed I have interpreted them, as fundamental building blocks of the religious imagination. But what about imagination itself? Can these apes be said to imagine? Do they pretend, or make up worlds for themselves? Does an evolutionary precedent exist for the young child's tea party, achieved with no tea, no cups, no saucers, and no attendees except those in the child's mind?

My favorite examples of ape imagination involve chimpanzees, one home-reared and one wild. In the 1940s, the chimpanzee Viki was raised by Keith and Cathy Hayes, who tried to teach her to talk. The experiment pretty much failed, though after extensive practice, Viki could make four wordlike utterances. And it seems that she could use her imagination:

Viki's favorite playthings were pull toys and picture books . . . one photo shows Viki putting her ear to a picture of a watch. Perhaps her most famous instance of pretense was her invention of an imaginary pull toy which she "pulled" through the house by an imaginary string. When asked what she was doing . . . Viki stopped short "with a look of guilt and embarrassment" and then "pretended to be very busy examining a knob" on the toilet. Over the course of a month or so, Viki continued to play with the imaginary pull toy, even to the point of its getting "stuck" and having her adoptive mother "untangle" it.[26]

What an extraordinary series of observations! Viki made up, in her head, a toy to amuse herself in a way instantly recognizable to anyone who has spent time with a child. This is true, of course, only insofar as the Hayeses' account is reliable. No reason exists to dismiss it, if we may judge by their careful reporting of their failures as well as their successes in training Viki. Further, Viki's capacity for imagination is by no means exceptional, either for enculturated apes or for "regular" captive apes (skeptics may consult Robert W. Mitchell's volume *Pretending and Imagination in Animals and Children*).

Let's dip back into my favorite-anecdote grab-box for a second example, this one from primatologist Richard Wrangham. Wrangham's observation of Kakama, an eight-year-old chimpanzee living in Uganda, shows that apes' imaginative powers are not produced by (though they may be encouraged by) their living with humans. Watched by Wrangham, Kakama carried around a small log, and retrieved it when it fell from a tree. Further, he fashioned a sleeping nest for the log, smaller in size but otherwise like those all chimpanzees build at night. "It looked like a toy nest," writes Wrangham. "After making it, [Kakama] first put the log in it. He sat next to it for two minutes before climbing in himself, rather awkwardly because the nest was small."[27]

Intriguingly, Kakama's behaviors with the log occurred when his

mother was pregnant and slept a lot. Perhaps Kakama felt a little lonely. He had shown, earlier in life, a particular propensity for spending time with younger chimpanzees. Knowing this, Wrangham concluded that when no younger companion was around, Kakama invented one.

Perhaps it is time to *search* for imagination in the African apes. Researchers can create the experimental conditions in captivity, and look with searching eyes in the wild. Surely the human imagination can dream up ways to study the evolution of imagination!

It may be the writer Edward P. Jones creating a compelling fictional universe in *The Known World*, or the musician Bruce Springsteen composing, and playing in concert with his band, a heartbreaking song about love and loss after September 11, 2001. It may be an astronomer interpreting pictures of deep space sent back by the Hubble telescope, or a doctor considering how best to design a clinical trial for a promising new drug to conquer AIDS. Or it may be a toddler creating a busy working farm in her bedroom, helped only by stuffed toys. All our most soaring feats depend on the ability to think beyond the here and now, to create a world in our heads that exists nowhere else. Humanity's relationship with the sacred is no exception. To imagine the beauty and depth of the sacred around us is not only to make meaning, but also to grasp the depths of that meaning. How powerful it is to realize that, although the human imagination makes our species unique, we are nonetheless connected to our closest living relatives, the African apes, in this way as in so many others.

CONSCIOUSNESS

What is the nature of human consciousness? Are some animals conscious? Which ones? Are there degrees of consciousness? Questions like these make consciousness a singularly contested issue in today's scientific world. In *A Mind So Rare*, the psychologist Merlin Donald clearly outlines the terms of the debate. The "hardliners," as Donald dubs them, insist that consciousness is a weak force at best and that mental life is largely driven by unconscious forces. For hardliners, the

human mind is controlled by deep-seated mental modules of a kind that I discuss later in the book. We may think we are making conscious decisions moment by moment, but this is all illusion. In reality, humans possess "a huge reservoir of unconscious or automatic cognitive processes that provide a background setting within which we can find meaning in experience. By relying on these deep automaticities, we can achieve great things intellectually."[28]

Even while recognizing the vast power of the hardliners' ideas in science today, Donald rejects these premises. He believes consciousness to be the primary force in our mental life, and convincingly argues that it is constructed by culture. Let's jump on Donald's bandwagon and leave the hardliners behind. This means working from a position that accepts the centrality of consciousness in human life. What about the apes? Do they show evidence of what Donald Griffin has called "the subjective state of feeling or thinking about objects and events"? That is, are apes conscious?

My answer is yes. The apes tell us so. Could the chimpanzee Flint have so grieved for his mother without *feeling* her death? Could the gorilla Binti Jua have rescued the unconscious boy without *feeling* or *thinking about* his vulnerable state? Could Ivory Coast chimpanzee mothers guide their youngsters' nut-cracking skills without *thinking* about how to go about it in appropriate ways?

The morning I began to draft this section, the *Chicago Tribune* published a moving article.[29] At the Brookfield Zoo, a thirty-year-old female gorilla with end-stage kidney disease was euthanized. Babs had been the group's matriarch, the high-status female in an extended social family that also included the famous Binti Jua. After her death, Babs's groupmates were allowed to view and touch her body. Babs's nine-year-old daughter held her dead mother's hand and stroked her stomach; other gorillas gently touched her body as well.

This account reminded me of the vignette in the Tai forest. The chimpanzee Tina has died from a leopard bite, and her little brother and other group members behave at the body in emotionally similar ways. The events surrounding Tina's and Babs's deaths, together with all the other evidence for ape socioemotionality, point clearly to the conclusion that apes have consciousness. Skeptics could, of course, argue. Perhaps Tina's and Babs's ape associates just sensed "something different" about an inert body, and investigated it out of curiosity or confusion. I would reply that in each case, relatives of the dead ape expressed emotions that seemed wholly unlike either curiosity or confusion.

But let's not forget the consciousness hardliners. What a position ape researchers find ourselves in! We are faced not only with the usual suspects, who argue that consciousness (or language or thought or culture) is uniquely human, but also the hardliners, who insist that human consciousness is weak even in humans, for we are so much controlled by our unconscious mind!

I cannot risk diverting this book, which is about religion rather than consciousness, down the wrong path. So I will leave the topic of consciousness with a comment. Surely the story of those who seek God, gods, and spirits, of creatures who turn their feelings of empathy into compassionate acts carried out in the name of faith, is the story of conscious beings. And just as our empathy, our meaning-making, our rule-following, and our imagination all have roots in ape abilities, so I believe does our consciousness. In prehistory, millions of years intervened between an apelike ancestral consciousness and our own—between apes who express empathy for a hurt friend, and Mother Teresa who devoted her life to compassionate works in honor of God; between apes who create imaginary toys, and architects who build magnificent cathedrals to be filled with the voices of all those who sing and speak the glory of the sacred.

What happened during those millions of years? How did the need for belongingness manifest itself in our extinct ancestors during that time span? That journey begins in the next chapter.

THREE

African Origins

APES EMPATHIZE with their friends and relatives, and make meaning with them. Apes follow social rules and use their imaginations. The previous chapter painted a picture in which humans share the earth today with apes who fit this description. Scientists know that humans evolved gradually, and shared a long evolutionary history with the apes. But how gradual is "gradual," and how long is "long"? To what degree are today's chimpanzees, bonobos, and gorillas good guides to the very first stages of human evolution? Most important, what changes can we describe, related to the evolution of belongingness and the origins of religion, once the human ancestral lineage began to diverge from the ape lineage?

This chapter acts as a bridge in tackling questions like these. Focusing on the time period between 7 million and $2^{1}/_{2}$ million years ago, it links the ancient ancestors of today's apes with our ancestors who, more recently than that, began to interpret their world through symbols. Little can be coaxed from the archaeological record about behavioral evolution before the material, visible engagement with symbols.

With some surprising exceptions to whet our appetite, almost no clues exist early on to the expression of emotion in daily life. It's thus best to admit right off that no cache of information about the origins of religion exists for this most ancient era of human evolution.

Yet, that era deserves exploration nonetheless, for it has a story to tell, an important story with twists and turns of its own. The origins of the religious imagination can only be fully grasped by looking at the entire sweep of the earth's history and our place in it; we can understand what came later by understanding what came first.

PEERING INTO OUR PAST

Every year in my house on October 23, we celebrate my daughter's birthday. Some households may commemorate a different event on this same date. In the seventeenth century, the Protestant bishop James Ussher famously calculated that God created the earth on the twenty-third day of October in the year 4004 B.C., or about six thousand years before our present day.

No person living in Ussher's time had any reason to entertain notions of an older earth. Historical records indicate that scientists, theologians, and artists alike accepted a date akin to Ussher's, though perhaps less precisely calculated. The astronomer Johannes Kepler believed 3992 B.C. to be accurate, while Martin Luther adopted 4000 B.C., as did Shakespeare when he had the character Rosalind say, in *As You Like It*, "The poor world is almost six thousand years old." How stunned would any of these great men be to learn that the earth is in fact 4.6 billion years of age? Would they feel awe if they could visit the Grand Canyon's South Rim, and feel with their own hands the weight and warmth of rocks (Vishnu Schist) nearly half that age?

For humans today, learning the ancient nature of the earth has special meaning, akin to learning a startling fact about the apartment building or house we live in or the garden we tend. The earth is our home; daily, we work its soil or swim in its waters or tread its ground. But of course, our planet's origin represents just a single, tiny point on the grand evolutionary time line. Learning to extract useful knowledge from this time line is like learning to use a camera to capture an event as it unfolds: we must learn what magnification to use and how to operate the lens in order to achieve the best results.

Cosmologists may zero in on the origins of the universe itself (about 13.7 billion years ago) and biologists on the origins of life on earth (about 3.8 billion years ago). The lens best suited for biological anthropologists focuses on a 70-million-year span of time. Around 70 million years ago, primates first emerged, then branched out in a series of "adaptive radiations" that continued straight through the appearance of anatomically modern humans about 200,000 years ago. Of course, human evolution does not stop at that point. Human beings 200,000 years ago walked like us and looked like us, but they were far from behaviorally modern, and they continued to evolve in a host of ways, as indeed we do today.

How amazing to stop and think that these four periods of origins— of the universe, of the earth, of life on earth, and of primates—connect to each other. As the astronomer Carl Sagan was fond of saying, we humans have "star stuff" in us: the chemical composition of our bodies reflects the makeup of stars. At the universe's birth, three simple atoms existed (hydrogen, helium, and lithium). Only when stars themselves were born did so-called star factories flood the universe with the diversity of elements around us—and in us—today, such as carbon.

Further, life on earth had a single origin, so that each of us is connected with every other living creature. "As humans," remarks Bill Bryson, "we are mere increments—each of us a musty archive of adjustments, adaptations, modifications, and providential tinkerings stretching back 3.8 billion years. Remarkably, we are even quite closely related to fruit and vegetables. About half the chemical functions that take place in a banana are fundamentally the same as the chemical functions that take place in you."[1]

With these connections firmly in mind, we can return to our own evolutionary journey as primates. Certainly, this lineage's 70-million-year history represents a period incomparably longer than Ussher,

Kepler, Luther, or Shakespeare could have imagined for *any* event in the universe. The fossil record shows beyond doubt that, somewhere between 80 million and 65 million years ago—with a wide variety of mammals already thriving and the dinosaurs on the very verge of extinction—a new kind of creature appeared.

In first developing from ancestral mammalian stock, primates exhibited early versions of the traits that would develop fully over the next millennia and that are still found in us today. Though anthropologists are fond of devising lengthy lists of primate characteristics, we can boil down to a quartet the most critical of these: grasping hands, forward-facing eyes, large brains, and slow growth with lots of learning during the infant and juvenile stages of life.

The next time you hammer a nail into a wall or knit a sweater, log on to the computer to seek a bargain on eBay, play the piano, or put a cool compress on the feverish forehead of a child, thank the ancestral primates. It is their legacy that led to the grasping hand that permits you to do these things, to grip objects and move your fingers independently and skillfully. Unlike most mammals, primates use their hands and fingers to eat: no monkey or ape equivalent exists to the domestic cat's face-dive into a bowl of Tender Vittles or the horse's muzzle-plunge into a bag of oats. As I observed baboons do in Kenya, monkeys and apes pluck fruits or blossoms off a tree, and dig up underground tubers or grasses from the earth. Some apes even modify wand tools that they stick down into the earth, or into holes in trees, to probe for delicious protein snacks in the form of insects.

But manual skills in feeding are only the half of it. Grasping hands allow a newborn to cling to the fur of his mother as she runs to keep up with her group or climbs a tree to find succulent blossoms. They keep mother and baby tethered, more like a single unit than two individuals, during the baby's first weeks and months of life. They underwrite grooming, the glue of primate social life in which social partners pick through each other's fur in a ritual expression of trust and closeness (and secondarily, a hygienic cleansing of dirt and bugs).

Forward-facing eyes combine with grasping hands to allow for depth perception, a critical ability for animals that spend much, and in some cases all, of their time in the trees. In the Amazon Basin's forest canopy live monkeys, ranging from tiny tamarins to bigger, aptly named howlers, who jump from branch to branch in search of food and shelter and who rarely if ever descend from the heights. On mountainous slopes in East Africa, and in thick forests in Asia, ponderously weighty apes climb branch to branch, fashion comfortable nests from surrounding vegetation, then settle down to sleep. Their binocular vision allows these monkeys and apes to make good judgments about when and where to make daring leaps or cautious creeps through the trees.

Primates seek both to eat other animals and to avoid becoming eaten themselves. Sharp vision, then, is directly tied to their survival. The very reason that the combination of grasping hand and forward-facing eyes evolved, at the dawn of the primate lineage, may involve the opening up of a new feeding niche. If the earliest primates encountered insects up in the trees, good hand-eye coordination would have allowed some individuals to capture and eat more protein-rich bugs than others, leading in turn to higher rates of successful reproduction for those with more highly developed grasping abilities and greater depth perception.

When primates first organized themselves in social groups, the reason was probably not just general "safety in numbers" but also, more specifically, safety in many pairs of forward-facing, depth-perceiving eyes. When many monkeys are able to orient with acute vision in multiple directions, the chances increase that a hungry leopard or eagle can be sighted before one of the group becomes its dinner.

Brains matter, too. The brains of primates are nearly twice as large in proportion to body size as the brains of typical mammals. Brain size increases again in humans, exceeding predictions for a primate of our size. My favorite "gee whiz" fact to cite in introductory anthropology class is not, however, our large brain size itself, but a consequence of it. A lengthy pregnancy, twenty-one months instead of a mere nine, would be required to produce human babies with brains as well developed at birth as those of typical mammalian infants. Though the prospect of a pregnancy lasting almost two years may fail to impress eighteen-year-old anthropology students, it sears the imagination of any woman who has birthed a baby. Evolution avoided this extreme solution; what happens instead is that the baby appears in the world with a uniquely "unfinished" brain, a brain that then grows at an astonishing rate during the first year of life.

Finally, because infant and juvenile primates are so slow-growing, reaching puberty (and the reproductive years) relatively late, they spend many years deploying their impressive brains as sponges. They soak up knowledge day by day from their elders, as I can attest from first-hand observations of infant and juvenile monkeys in Kenya, and of apes in a variety of captive settings. Primate immatures follow, watch, and, in some cases, copy their elders. More than passive observers, they actively work to construct situations in which they can learn about every aspect of their lives: not only which foods to eat and which other animals to avoid as enemies, but also which other group members are congenial and which aggressive, when to approach dominant males and when to avoid them, and how to play exuberantly or express fear in appropriate ways.

These four features—grasping hands, forward-facing eyes, big brains, and a lengthy period of learning—sum to a whole that is more powerful than its parts. They make for a very special kind of mammal, one highly adapted to long periods of learning aided by brainpower, visual skills, and manipulative capacities. To apply the evolutionary perspective to maximal effectiveness, however, we need to look beyond a generic, all-purpose type of monkey or ape. Over two hundred primate species are alive today. As we learned in the previous chapter, some are much more relevant than others to our consideration of an evolutionary underpinning for the human religious impulse, that is, of behavior rooted in belongingness and emotional connection. Where do the empathetic, meaning-making, rule-following, imagining, conscious apes fit in to the picture?

Commonly, primates are divided into two major groups, the

prosimians ("premonkeys") and the anthropoids ("like humans").[3] Prosimians evolved first, soon after our 70-million-year starting point, from ancestral mammalian stock. Today they include the nocturnal slow loris of India and Sri Lanka, and the ring-tailed lemur from Madagascar, off the East African coast. While lorises and lemurs exhibit fairly complex social patterns, it is nonetheless safe to conclude that prosimians are less fully developed in terms of social behavior and intelligence than are the anthropoids.

For many people, an anthropoid is an anthropoid is an anthropoid. One of the more interesting experiences of doing zoo research is watching parents who tug their children into a building, directly under a large sign proclaiming "Great Ape House," only to point out, to their own little clinging primates, "the big monkeys!" And how often on television shows or in newspaper articles is a clip or photo of a gorilla or a chimpanzee captioned "Monkey Business"? To a primatologist, such conflation of monkeys with apes approaches the scandalous. A quick and dirty test within the anthropoid category centers on the tail. Monkeys have tails; apes do not. But of course, this is only a relatively superficial marker. More important is that of the two groups, monkeys evolved first. Another way to say this is that apes evolved closer in time to the birth of the human lineage itself. Whether we choose as a yardstick the number of chromosomes, the skeletal arrangement of muscles and bones, or various aspects of behavioral complexity, apes are more like us than monkeys are. The difference is particularly pronounced when we zero in on intelligence. No matter whether it is assessed through analysis of general problem-solving or specific patterns of tool-using, symbol-using, and expression of empathy and imagination, apes' intelligence outranks monkeys'.

For all these reasons, apes are the gold standard for anthropologists who seek to explain humanity's past. But just *how* can apes help us understand prehistory? The answer is not entirely straightforward, but certain myths can be dispensed with straightaway. Humans did not evolve "from apes." Never mind all the versions of the popular drawing or cartoon that links them, left to right, starting with a hunched and hairy chimpanzee, next a stooped, club-wielding, and none-too-smart-looking human ancestor, then a slightly more present-able "caveman" type wearing an animal skin. Last, on the right, proud and erect, stands a spear-carrying, highbrowed modern man (yes, these drawings always *do* depict males). Used to illustrate everything from articles in popular science publications to print ads for a telephone company in the United States to billboards in the London under-ground, this supposed evolutionary progression is about as misleading as it could possibly be. No linearity, and no direct ancestry, character-izes the relationship between apes and humans.

Equally wrong is the idea of a "missing link." No single fossil repre-sents the key bridge on the evolutionary line or plugs the gap between primitive ape and modern human. The term "missing link" makes for the basis of a nice sound bite. Using it to describe a key fossil find is an overwhelming temptation for some, including those who should know better. Even the writer Maitland Edey, who teamed with the anthro-pologist Don Johanson to write a book about the famous fossil "Lucy" found in Ethiopia, earlier published a book called *The Missing Link*, focusing on a group of early human ancestors called the australo-pithecines. The book's title implies that these prehumans link Jane Goodall's chimpanzees in Africa with the human inhabitants of New York or Nairobi today.

In some ways, the anatomy and behavior of australopithecines may *appear to us* to fall between that of apes and humans. In reality, though, their physical and behavioral characteristics should be understood as adaptations to the environmental conditions of their own time and place. Like all human ancestors, australopithecines lived and died not as links between what came before and what came after, but as flesh-and-blood creatures with day-to-day concerns ranging from getting enough to eat, to relating emotionally in some way to their social partners.

Linearity simply won't do as a way of thinking about human evo-lution. Far less catchy than "the missing link," or even a tree of origins, but having the virtue of accuracy, is the concept of the common ances-tor. *Humans did not evolve from apes; apes and humans shared a common an-*

cestor. This last sentence is crucial to the evolutionary perspective; every fall in my introductory class, I ritually chalk it on the board in large letters. To grasp that humans and apes shared a single origin point and then split into two lineages, one leading to modern apes and the other to modern humans, is a prerequisite for a full understanding of the evolutionary perspective.

Of the various kinds of apes, the chimpanzees, bonobos, and gorillas of Africa are, of course, our closest cousins. About 8 million years ago, none of these great apes yet existed, nor did humans or their ancestors. Alive back then was a generalized apelike creature that had itself developed from earlier, monkeylike forms. The roots of this apelike creature can be traced back 70 million years, when primates, with their quartet of key features, diverged from ancestral mammals. Slowly over time, this generalized animal began to change in response to different environmental pressures. Evolutionary "radiations" from the single point occurred in waves.

Precise laboratory techniques for the study of DNA have produced key discoveries in the chronology of human evolution. First, humans are most closely related to chimpanzees and bonobos, and less closely related to gorillas. This means that references to *the* common ancestor oversimplify evolutionary accounts. Multiple "splits" or evolutionary divergences occurred over the last 70 million years, so multiple common ancestors must have existed; most recent among them was the human-chimpanzee-bonobo common ancestor. Second, the molecules tell us about absolute as well as relative timing. The African apes (including gorillas) had a common ancestor with humans at around 8 million years ago, whereas chimpanzees and bonobos alone had such an ancestor around 6 million–7 million years ago.

However, it's a little disingenuous to present this account as if all these dates, stemming from analysis in the laboratory, mesh perfectly with what is known from digging up fossils. As I discuss below, fossil hunters recently discovered a human ancestor that lived in Africa right around 6 million–7 million years ago, a finding that crowds the predicted split date for the chimpanzee and bonobo lineage from the human one. There wasn't supposed to *be* any hominid quite as early as that! Further, it's an understatement to admit that not everyone accepts the accuracy of the laboratory techniques that lead to these dates. (The lab scientists shoot back that dating of fossils isn't exactly a precision science, either—and on goes the debate.)

Beyond doubt, however, are two facts. Humans' closest cousins are the African apes—chimpanzees and bonobos first, then gorillas. Next most closely related to humans are the apes of Asia (including the orangutan), and after that, the monkeys. As a result, the most useful models for the hominids who lived after the great evolutionary divergence at about 7 million years ago—a time period for which we have scant physical remains, and even scantier insight into behavior—are the African apes.

What can we conclude from peering into our distant past? *The ways that chimpanzees, bonobos, and gorillas experience belongingness and express emotion give us our best clues to how our earliest hominid ancestors, alive right after the ape-hominid split, led their lives.* Some anthropologists would be quick to point out the risk inherent in a claim of this sort. Perhaps the African apes developed their abilities to empathize, to negotiate meaning, to follow rules, to use their imaginations—indeed, to be conscious—not before the ape-hominid split but afterward, at some subsequent point in their own evolution. If this was the case, the African apes would be poor models for the earliest human ancestors. Yet I think the risk is minimal, and acceptable. It's a good bet that the cognitive and emotional behaviors I've been describing for the African apes were fundamental characteristics of the ancient ape lineage, because they are expressed so robustly in the various species of apes under so many different environmental conditions today. Though not unassailable, my conclusion is reasonable. Where does it take us next?

OF BUSHES AND TREES

All humans began as Africans.

The ancestral homeland for all of us, from the Inuit in northern Canada to the Israelis and Palestinians in the Middle East, from rural southerners in the United States to urban Chinese and Chileans, is Africa. Charles Darwin strongly suspected this fact in the nineteenth century, and when I lecture around the country on human origins, I find that most people are aware of it now. Some reject the concept that humans evolved at all, believing instead that God created humans, starting with Adam and Eve in the Garden (see Chapter 8). Still, most people, no matter their beliefs, can conjure up images made famous by the study of African prehistory: the Leakey family scrambling over fossil bones in Olduvai Gorge, Tanzania, or Don Johanson's team piecing together the Lucy skeleton at Hadar, Ethiopia.

But with new discoveries, the human family tree is constantly being redrawn; even dedicated experts strain to keep up. Try out this query on a family member or friend (no fair asking a paleoanthropologist or strict creationist!): Where and when did the first hominids (human ancestors) live? Try to get the person to reply in more detail than "Africa" and "millions of years ago."

I suspect that few of you will hear the reply "The Sahara Desert, specifically the country of Chad in central Africa, at six or seven million years ago." The discovery in arid Chad of what seems to be the oldest hominid anywhere on the planet, took scientists by surprise and hasn't yet seeped into the popular consciousness. Bits of hominid material had been found in that part of the world before, but nothing so spectacular, or so ancient, as this. The evolutionary action was thought to have taken place about 1500 miles to the east.

There, in East Africa, is the 6,000-mile-long gash in the earth known as the Great Rift Valley. Running from Lebanon in the north to Mozambique in the south, the valley was formed by violent upheaval beneath the earth and remains geologically hyperactive today. It encompasses Olduvai Gorge and countless other hominid sites. Living in

Kenya, where the valley runs north–south and bisects the country, I never tired of wondering at the hominid fossils and living sites that surrounded me: some famous for successful excavations that shaped notions of our prehistory; some waiting to be discovered by future scientists; and others that will remain buried, invisible to the human eye. These places, together with numerous cave sites in South Africa, are part of our earliest heritage and tell us in no uncertain terms that Africa is our homeland.

But the *Sahara desert*? What in the world were we doing there? Evolving, apparently. Of course, at 6 million or 7 million years ago, this area of Africa had an entirely different ecology. There, in woodland habitat, the hominid lineage got its start—at least, as far as we know right now. In 2001, Michel Brunet and his colleagues discovered a skull, fragments of a jaw, and several teeth that, in combination, look like no known hominid. Whereas the skull is the size of a chimpanzee's, robustly built with ridges over the brows and a crest on the top, the face is fairly vertical, whereas in an ape's face the jaw is thrust out in front of the skull. The sharp canines of an ape are absent. This odd mosaic of features adds up to a hominid so unlike any other as to be anointed with a new genus, *Sahelanthropus*, named after the Sahel region of the Sahara, as well as a new species, *tchadensis*, named for Chad.

As the paleoanthropologist Bernard Wood has said about *Sahelanthropus*: "A single fossil can fundamentally change the way we reconstruct the tree of life."[4] Lessons learned from this one are both geographical and temporal. In an interview with the BBC after the Chad find, Lucy's discoverer Don Johanson commented: "If we could open windows—geological windows into other areas—we would see that hominids were much more widely dispersed than previously thought."[5] Much older than previously thought, too, he could have added. When Johanson discovered Lucy, and dated her to 3.2 million years, *she* was the oldest hominid known.

Some anthropologists, it should be said, insist that *Sahelanthropus* is not a hominid at all. They conclude, on the basis of their considerable expertise with bones, that this fossil represents one form of an ape-human common ancestor. The skeptics are in the minority, however; most experts believe that *Sahelanthropus* significantly extends our family tree both back in time and across space.

"Family tree," though, is one of those seductive terms: everyone knows what it means, so it is easy to use. But it misleads. Trees are rather linear in shape, growing more vertically than, say, bushes. More and more, we are finding that the story of human evolution is "bushy," characterized by multiple forms living at the same time and not one form succeeding another in a linear way.

Hominid prehistory, then, begins 6 million or 7 million years ago. As it unfolds, only rarely does one distinct species cleanly replace an earlier distinct species. More often, multiple species coexist, with some lineages becoming extinct while others leave descendants. Technically, "prehistory" ends when writing begins. This book, however, concerns itself with the evolution of the religious imagination only up until the time of agriculture. Only very briefly do I touch on events that occurred once human groups began to domesticate crops and to settle into communities.

The trick is to convey something about the diversity of hominids during prehistory without bogging down, fossil by fossil, in technical details. A glance at the most likely current evolutionary time line is convincing enough: for a full grasp of hominid anatomy and behavior, one needs a course or text designed to convey detailed information. Happily, excellent sources are available for those who wish to delve more deeply into this material, and I must make clear that I do not aim for a comprehensive account here.[6] My goal is to describe general patterns in human evolution in order to assess what can be inferred about the origins of religion. This chapter concerns only the earliest hominids living in Africa, divided into three groups: those living before 4 million years ago; australopithecine hominids living after 4 million years ago, including Lucy; and the first hominids in our own genus, *Homo*, appearing at about $2^1/_2$ million years ago.

BEGINNINGS

Australopithecus afarensis, made famous by the Lucy fossil, is a touch-stone species in both the scholarly and public understanding of human origins. Fully 40 percent of Lucy's bones were found, with material from both the cranial (head) and postcranial (below the neck) areas. When Johanson announced to the world that Lucy lived 3.2 million years ago, he rocked the world of science. The very root of our lineage had been discovered!

Undeniably, Lucy walked as a biped. This is known to a certainty from the anatomy of her knee and hip as well as certain features of her skull. Yet her brain was no larger than an ape's. Upright walking, Johanson suggested, crucially distinguishes hominids from apes right from the first. In the mid-1970s, then, bipedalism was christened the hallmark feature of those species who had crossed the Great Divide—that is, of species on "our" side of the ape/hominid split.

Though paleoanthropologists hotly debate exactly how Lucy walked upright, the consensus about her bipedalism has never wavered. Yet thanks to *Sahelanthropus*, the known span of human evolution has nearly doubled. Were the earliest hominids, those living between 7 million and 4 million years ago, also bipedal? Is bipedalism truly the watershed between other primates and human ancestors? What *do* we know about the earliest beginnings of our own lineage, and can any of that knowledge inform questions about the origins of the religious imagination?

Let's revisit the current contender for Lucy's former title, "oldest hominid." *Sahelanthropus* is also known as Toumai, "Hope of Life," in the local African language. Nicknaming fossils is a common practice, maybe because it's easy to develop a fondness for our early ancestors. At once so familiar and yet so strange to us, these fossils "connect us to eternity," writes Glenn Conroy, "giving us the privilege of glimpsing, if ever so briefly and imperfectly, the blurred image of those who walked the earth" in the distant past.[7]

Grouping fossils for discussion highlights broad patterns of adaptation. When I treat a single species at length, I introduce the technical Latin binomial (such as *Sahelanthropus tchadensis*, *Homo habilis*, or *Homo sapiens*) in addition to any nicknames. The scientific terms convey useful information. They either clue us in to the fossil's location ("Sahel"; "Chad"), or tell us something about prevailing assumptions at the time of naming (*habilis* means "handyman"; *sapiens* means "wise"). Names hint at the inferred relationships among hominids. *Homo habilis* and *Homo sapiens* are placed in the same genus because they are thought to be more closely related to each other than either is to *Sahelanthropus*.

Only cranial material has been found for *Sahelanthropus*: no hip, knee, leg, or arm bones reveal how this creature moved from place to place in its ancient world. Did these hominids knuckle-walk on the ground, and swing beneath branches when in the trees, as today's chimpanzees and gorillas do? Did they walk upright, even if in a primitive way? Perhaps they locomoted in a unique way? The position of the skull's foramen magnum, the hole through which the spinal cord passes, hints that *Sahelanthropus's* posture may have been relatively apelike, with the head thrust forward rather than balanced vertically atop the body. Other experts see traits that point to bipedalism. The jury is still out on this one.

Postcranial bones *have* been found for other very early hominids. And lo and behold, these hominids were found where they were "supposed" to be: East Africa. In both Kenya and Ethiopia, fossil material dated prior to 4 million years ago represents creatures that look somewhat chimplike, and somewhat humanlike. Leg bones are a major prize because they figure so prominently in the race to pinpoint the antiquity of upright walking. Of course, we can dispense right away with the rosy image of a roomful of paleoanthropologists nodding in collegial agreement as they scrutinize such bones. A case in point is *Orrorin tugenesis*, found in the very heart of Kenya's Great Rift Valley.

Dubbed Millennium Man, for the year of its discovery, *Orrorin* is solidly dated to 6 million years ago—old enough for this fossil to have reigned briefly, before its dethronement by *Sahelanthropus*, as the oldest hominid. The scientists who uncovered *Orrorin* interpreted aspects of its femur, the long bone in the thigh, as evidence of the kind of strongly muscled leg that marks a biped. Skeptics wished for independent corroboration. In 2004, support for Millennium Man's bipedalism emerged not from the bones themselves, but from computerized tomography. CT scans of *Orrorin's* bones show that the precise relationship between the thicker and the thinner parts of the femur neck is a strong indicator of erect posture. Bipedalism is more ancient than Lucy: that much is clear.

Fossils from Ethiopia ranging back to 5.8 million years ago, nearly *Orrorin's* time, represent yet another early hominid. With *Ardipithecus*, too, tantalizing glimmers of bipedalism exist—and perhaps more than glimmers. Paleoanthropology is a swiftly changing field. The very week I was writing this section, a set of new fossil finds from Ethiopia was announced. Bones of nine hominid individuals assigned to *Ardipithecus* were uncovered, including a foot bone with features "diagnostic of bipedality," as the formal report puts it. Informally, speaking to the press, the archaeologist Sileshi Semaw declared that the foot bone confirms hominid bipedality in *Ardipithecus* at 4.5 million years ago.[8]

At this early time period, even taking into account the new discoveries, scientists are working largely from fragments and scraps of bone. Paleoanthropologists, staking whole careers on these fossils, become intensely passionate about the shape, form, and function of a thigh bone there, a toe bone here. They defend their turf with vigor, insisting either that these hominids were bipedal or that they were not. Compared to these bits of bone, Lucy's 40-percent-complete skeleton is a gold mine.

What can be concluded about this earliest time period of our lineage? First, the human bipedal adaptation is likely, if arguably, as ancient as our lineage itself. Second, analysis of these early hominids is, literally, close to the bone (and the muscle). No reliable way exists to explore their social behavior, or tool-making behavior, or indeed any behavior at all beyond walking and eating. What types of groups they lived in remains a complete mystery.

Further, much is up for grabs in our understanding of the earliest human ancestry. Comparison of the various species living before 4 million years ago tells us that our lineage was launched without any singularly successful anatomical blueprint, indeed without any single "Garden of Eden" location for the early incubation of our species. *Sahelanthropus, Orrorin,* and *Ardipithecus* all mix more apelike and more humanlike features, but in distinctly different ways, and in different parts of Africa. That woodland habitat plays such a prominent role at the dawn of our lineage is a fairly recent revelation, and may hint at a real and significant pattern.

ABOUT RELIGION

There's no use trying to avoid a single stark fact: the fossil record at this earliest time period reveals nothing whatsoever about the origins of religion. While it is reasonable to assume that the first hominids had cognitive powers, and emotional capabilities, as developed as those of apes, the fossil record remains completely silent on this point. Eloquent about the length and diversity of our hominid prehistory, and full of anatomical hints about our ancient bipedalism, these creatures remain mute about how humans evolved God.

AUSTRALOPITHECINES

After about 4 million years ago, the fossil record becomes considerably "bushier." Students of our ancestry confront a veritable explosion of hominid names and site locations to deal with. Living in the woodlands and savannas of Africa were a number of hominids, more humanlike than *Sahelanthropus, Orrorin,* or *Ardipithecus* but not yet humanlike enough to be classified as *Homo.* These hominids can be broadly classified as australopithecines, members of the genus *Australopithecus.*

Whereas the ink is still drying on the initial announcements about the newfound hominids, australopithecines have been known and

studied for over eighty years. In the year 2024, anthropologists around
the world will commemorate the hundredth anniversary of australopithecine studies. In so doing, they will honor Raymond Dart.

A South African anatomist, Dart was brought a skull from a mine.
For more than a month, he laboriously chipped away the debris clinging to the skull, nicknamed Taung after a village near the mine. In
1925, Dart announced to the world, via the journal *Nature*, that the
skull represented a creature ancient (2 million years old), small-
brained, and bipedal. The clue to bipedality lay in the position of the
foramen magnum. Dart coined a new name for the Taung skull: *Australopithecus africanus*, "southern ape of Africa."

If we could climb into a time machine, activate the Reverse lever,
and hurtle back to the 1920s, we could grasp the revolutionary nature
of Dart's claims. His trifecta of proposals was lethal in its day—a human ancestor of 2 million years; an African origin for humanity, rather
than a European or even Asian one; and *small* brains, not big, clever
ones, at the root of our human ancestry. In one fell swoop, Dart had violated the scientific community's most cherished expectations about
our earliest ancestor.[9] To say that he received no accolades for his report on Taung would be an understatement; scientists who did not ignore his announcement scorned or rejected it.

Readers might suspect that, by now, scientists would have nailed
down answers to all the mysteries of *A. africanus*. But in its Hall of Ancestors website, the Smithsonian Institution calls it "an enigma," admitting that "researchers are still unsure about where *africanus* came from
and which species, if any, it led to."[10]

What *is* known about *africanus*: that the Taung skull comes from
the southernmost tip of *africanus*'s distribution, which ranged across
South and East Africa; that Dart was right to think these hominids
walked upright, though they may have climbed trees as well; that they
existed for perhaps a million years, indicating a long and successful
adaptation. Males were considerably larger than females; on average
males were under five feet tall and under a hundred pounds, but females were under four feet in height and twenty-five pounds lighter.

Here is an opportunity to look at key patterns of adaptation. Happily, there's compensation for the "bushy," pesky diversity of the australopithecine species. Fossils can be sorted into two broad groups, allowing a clearer look at the big picture. Into one group fall the gracile hominids. Comparatively light and slender in build, with no particular specializations of the skull or body, the graciles include both *Australopithecus africanus* and the famous Lucy, *Australopithecus afarensis*. Computerized animations (or museum dioramas) of early bipedal hominids trailing across the Ethiopian plain depict the gracile adaptation.

The robust hominids, by contrast, were bigger and bulkier. Their skulls were specialized, including, in some cases, a crest riding along the top, used to anchor massive jaw musculature. Teeth of *Australopithecus robustus* and *Australopithecus boisei* were adapted for crushing and grinding. For a long time, anthropologists concluded that the two australopithecine groups must have eaten radically different diets—hard nuts and seeds for the robusts, versus softer fruits and meats for the graciles. Newer testing has proven that robust hominid teeth bear the signature of meat-eating, a finding scientists don't yet fully understand.

Not every hominid living between 4 million and 1 million years ago neatly sorts into one of these two categories. Still, the gracile-robust dichotomy works well on a basic level. The two lineages experienced polar opposite fates, in the evolutionary sense. Some gracile australopithecines morphed into later hominids, leading eventually to modern humans, but the robust line became extinct. By about a million years ago, all traces of the robusts disappear from the fossil record. Most likely this happened because the environment shifted in a way incompatible with their specialized anatomy and (possibly specialized) diet. Before proclaiming them an evolutionary failure, though, best to think hard—the robust australopithecines flourished for nearly 1½ million years. Paltry by comparison, our own species' record extends little more than 200,000 years into the past. Only time will tell if *Homo sapiens* equals, much less exceeds, the lifespan of the robusts.

Televised documentaries routinely depict australopithecines foraging together, across the savanna-woodland, as a group of males, females, and children. In truth, we know nothing of their social units, whether for the gracile or the robust lineages. In research on australopithecine anatomy, technology, and rates of development, however, scientists are making surprising progress.

Reading the bones tells us that australopithecines strode bipedally some of the time, with hands free to carry infants or tools. At other times, they climbed trees, perhaps in search of shelter and safety. Their diet, too, seems to have been flexible rather than narrow, encompassing fruits, insects, and meat, though some of the robust species may have specialized in crushing and grinding.

Archaeology (in East and South Africa) and analogy (with living chimpanzees) tells us that some australopithecines probably made and used tools. Among the ranks of the first tool makers (described below) who processed animal carcasses for meat and marrow were probably some australopithecines. Further, computer analysis of striations on tools made of bone, dated to between 1.8 and 1 million years ago, shows that hominids in South Africa used bone tools to break open termite mounds.[11] Just think how different an image we would carry around of our past, if our ancestors were found to be insect-snackers as much as meat-eaters! Still, whether robust australopithecines or early *Homo* used these South African bone tools on termite mounds isn't yet perfectly clear.

Assessment of comparative growth patterns of teeth suggests that the period of immaturity in australopithecines was relatively short compared to ours. Apparently, the pattern of australopithecine dental development resembled that of chimpanzees rather than that of modern humans, at least in timing. Like chimpanzee babies, then, australopithecine infants probably matured more rapidly overall than our babies do. Here we move oh-so-close to the heart of the realm of belongingness, for the length of immaturity relates closely to learning and emotional relating across the generations. At the heart of this book, though, is the contention that ape emotional connections themselves are well developed, and these obviously are not precluded by a comparatively short maturation period.

More significant is the shift that comes from ape knuckle-walking and brachiation (swinging from branch to branch) to hominid bipedalism, and what that shift means for caretaking patterns. As bipedalism evolved, primate babies would have lost the ability to cling with grasping hands and feet to their mothers' bodies. Carrying an infant costs more energy than allowing it to cling. Early hominid mothers, as they walked and foraged, must have put their babies down at certain intervals, certainly more than ape mothers did. As the biological anthropologist Dean Falk has suggested, this change probably led to a cascade of others: hominid babies vocally protested at separation from their mothers, and their mothers in turn were selected by evolution to know how to soothe their babies, with both voice and gestures. The emotionally engaged "system" of mother and baby, particularly the empathy co-created by mother and baby together, would have become more exquisitely fine-tuned as bipedalism evolved in prehistory.[12]

To coax more than this about behavior and emotion from the australopithecine fossil record is difficult. One good story is left to tell, however, and it relates directly to the subject of this book.

COBBLING CLUES TOGETHER ABOUT RELIGION

It might seem highly improbable that a substantive debate could be engaged about the emotional or religious tendencies of australopithecines, yet a single discovery catapults us right into that territory. This is the Makapansgat cobble. Anyone who has joined with a child in finding animal or human faces in nature—in clouds or rocks—will understand the excitement generated by analysis of this piece of stone.

My description of the cobble is adapted from the report of the cognitive epistemologist Robert Bednarik.[13] Discovered in Makapansgat, a South African cave that also contained the remains of gracile australopithecines (*africanus*), the cobble is reddish-brown jasperite dated to almost 3 million years ago. Weighing about half a pound (260 grams), it features three depressions, located centrally on one of the two flattened surfaces. "Their striking appearance and distinctive arrangement

© R. G. Bednarik   5 cm

The Makapansgat cobble
Robert G. Bednarik

strongly convey the impression of a face," Bednarik writes.[14] Most intriguingly, the face one sees depends entirely on the cobble's orientation: look one way and a modern face is clearly visible; turn the cobble around, and the face that appears looks, at least to modern eyes, very much like that of an ancient hominid.[15]

No scholar, to my knowledge, has suggested that the facelike images were created by hominids. In fact, the cobble is completely unmodified from its original state in nature. What, then, makes it important to prehistory?

Most critically, the Makapansgat cobble is a manuport, an object carried in the hands (note the "manu," as in "manual"). In fact, archaeologists estimate it was carried into the cave, though whether by the australopithecines who used the cave or by other hominids in the area is not clear. The cobble could not have originated in the cave: analysis of the jasperite shows that at various times it "became the subject of fluvial action," meaning that it was carried along by flowing river water, and the cave shows no sign of water-transported sediments.

Paleoanthropologists have dismissed, based on analysis of the evidence, other ideas to explain how the cobble reached the cave. Could modern humans have transported the stone into the cave? Could it have arrived in the gut of a bird? In the end, the best explanation is the manuport one.

Two facts are key so far: the cobble bears an unusual resemblance to a face (either human or hominid, depending on one's perspective), and it was transported, at some trouble, into a cave used millions of years ago by australopithecines. The next question is obvious: is it *because* of the resemblance to a face that the stone was collected and kept by these hominids? Bednarik and the archaeologist Brian Hayden think so.

Bednarik notes that the cobble's markings are far too striking to have been ignored by creatures with the cognitive complexity of australopithecines. He dubs the Makapansgat cobble the earliest "paleoart." Even bolder is Hayden's linkage of the stone with self-awareness and with the possibility that the australopithecines' "curating" of the cobble might amount to "the first inklings of a notion of soul or afterlife" in prehistory.[16]

It is worth unpacking these views about the Makapansgat cobble for what they have to tell us about the search for our religious origins—and, indeed, about the nature of scientific inquiry more generally. Bednarik and Hayden both want to find evidence of unexpectedly complex qualities in early hominids. It is entirely natural, then, that they would be inclined to see hints in the cobble of self-awareness, art, or even spiritual inclinations. In our studies of this early period in human evolution, everything hinges on interpretation, and nothing outlandish mars Bednarik's or Hayden's ideas. Indeed, up to a point, I concur; it is probable that the Makapansgat hominids would have recognized the facelike features in the cobble, and might have been intrigued enough to collect, and even curate, the cobble as a result. Further, the existence of self-awareness in hominids of this time seems unsurprising; today's apes are conscious, so why shouldn't our ancestors of 3 million years ago have been conscious as well?

Nonetheless, suggestions of emerging spirituality—as distinguished from suggestions of self-awareness—based on the cobble make me uneasy. Hayden makes no definitive claims along these lines for the cobble, but neither does he acknowledge that collecting a stone with a human (or a hominid) face might amount only to a form of self-admiration or self-recognition. Perhaps the cobble was something pleasing or intriguing to gaze at, or to share with one's companions.

Let's imagine that some future primatologist stumbles across a group of chimpanzees who carry around a stone and keep it safe from harm. The stone has, at least in the eyes of the primatologist, apelike features. Surely, headlines would trumpet this discovery worldwide, and people would wonder whether the chimpanzees kept the stone because they recognized the face it bore. They would wonder whether the apes might be self-aware. Would people also wonder whether the apes were expressing a rudimentary form of art appreciation? Possibly. How about a rudimentary form of spirituality? I doubt it; after all, there's no evidence of spirituality in apes. And there's no evidence of spirituality in hominids at the Makapansgat-cobble time period, so why should we discuss the hominids any differently than we would a group of chimpanzees?

Caution is required when reading the past by reference to the present. Today, people may become quite excited if they believe they recognize a spiritual likeness in a natural object or an object made by humans. A fascinating list can be compiled of instances when this has happened, but a single case makes the point. In 1978, Maria Rubio, a woman living in New Mexico, was preparing a burrito. She noticed skillet burns on the tortilla that, to her, resembled the face of Jesus. Wishing to share her discovery with others, she opened her home to 8,000 people who viewed the image.*

Of course, no one has suggested that australopithecines saw the face of a beloved spiritual being in the Makapansgat cobble, but speculation of any kind that links the cobble with spirituality should be rejected. In sum, no evidence for any hominid before *Homo* points to an engagement with spiritual matters.

OUR OWN GENUS

Mild satisfaction? Outright exhilaration? Or no emotion, at least no emotion recognizable to us today? What would our ancestors have felt as they first used a sharp-edged tool of their own making to slice through a tough animal carcass? Imagine the scene: Ethiopia, about $2\frac{1}{2}$ million years ago. Here, as indeed across a wide swath of Africa,

*The story appeared in *Newsweek*, Aug. 14, 1978; see also Chapter 8.

hominids are vulnerable creatures. Living among predators equipped with stealthy bodies and sharp canines, they themselves are not naturally endowed to seek prey. Getting enough to eat from the land, while avoiding being eaten by roaming predators, must have been a considerable challenge. Skilled knapping of stone to make tools surely made a difference to hominid survival in this context. And, more to the point in the game of evolution, it would have aided reproduction, the ability to find a mate and live long enough to keep your lineage going strong.

The earliest known stone tools are dated to 2.6 million years ago. Found in the early 1990s at Gona, Ethiopia, they are relatively general-purpose in function. By analyzing marks on the stone, and by experimenting with stone knapping themselves, archaeologists reconstructed how the Gona tools were created: a comparatively softer "hammer stone" was struck against a harder, crystalline stone of basalt, quartz, or chert.

This percussive technique produces a sharp flake, with a core or cobble left behind. Flake tools can cut through animal carcasses or strip plant material, whereas the core can aid pounding and digging. In excavations at Gona, Michael Rogers hit the archaeological jackpot, finding flake tools together with animal bones of zebra and wildebeest— creatures that made up a healthy-sized dinner for Gona hominids.

Who were these hominids? Oddly, no hominid bones have yet been unearthed at Gona, so the identity of the tool makers remains a mystery. At a site called Bouri, nearby, is a different mix of evidence found and evidence absent. Antelope bones, $2^1/_2$ million years old, show distinct cutting by stone tools that can only mean hominids practiced butchery here, too. At Bouri, bones of australopithecines are in evidence but the tools are lacking.

Putting together the data from these two Ethiopian sites, we have definitive proof that at 2.6 million–2.5 million years ago, our ancestors were obtaining nutrient-rich meat and marrow by using stone tools of their own creation. Though the early percussive technique may sound to us like crude stone-bashing, archaeologists emphasize that it requires considerable skill. Analyzing Gona cores and flakes, the archaeologist Sileshi Semaw concludes that they were made by experienced knappers: "Ancestral tool makers chose appropriate size cobbles when making artifacts, selected for raw materials with good flaking quality, sought for acute angles when striking cobbles and produced sharp-edged implements used for cutting."[17] Here we begin to see why the Gona tools should be called the first *known* stone tools in hominid pre-history. They may be too elegant to represent the first-ever crafted stone tools, and as we will see, a simpler stone technology may have predated them.

Tools like the ones found at Gona (and inferred at Bouri) are la-beled Oldowan, after Tanzania's famous Olduvai Gorge where they are also found. Carrying replicas of Oldowan flakes and cores into my an-thropology classes, I ask my students to name some of their technolog-ical descendants. Answers come fast and furious: everything from the toothbrush to the Hubble telescope, the printing press to the laptop used by the students even as I lecture (used, I hope, to take notes and not to instant-message their friends). Oldowan tools are the first tan-gibles in the story of human cultural and technological evolution. From this point on, we can trace the step-by-step development of technology, from the next stage in African prehistory straight through to the latest NASA rocket. After Gona, technological development is written into the archaeological and historical records for us to read.

But as always, an evolutionary perspective compels us to look be-fore as well as after an apparent hominid turning point. Did apelike an-cestors make and use tools even before the ape/hominid split? The use of stone tools intentionally to make other stone tools seems to be a genuine hominid novelty; no chimpanzees or other nonhuman pri-mates do this today, as far as we know. Chimpanzees, are, however, for-midable tool makers and tool users in behavioral arenas ranging from feeding to grooming. In some places, chimpanzees crack open hard-shelled nuts with stone hammers using a technique so precise it takes about eight years to fully master. Indeed, chimpanzees in every studied population, without exception, make and use tools of some sort. This enormous variety has now been extensively catalogued, with many differences between populations attributable to culture rather than to genes or to ecology.[18]

Chimpanzees are highly unusual in the animal world for the extent of their commitment to technology. As the anthropologist William McGrew has noted, chimpanzees are far from being "one-shot wonders," like sea otters, who bang open mollusks to eat against stone anvils held on their chests but use no other tools: "Material culture is part of every aspect of chimpanzee life: subsistence, sociality, and self-maintenance."[19]

Chimpanzees are highly unusual among African apes in this respect, too. Bonobos do not make or use tools, and gorillas are known to do so only rarely. One population of orangutans (the Asian great ape) made and used tools, but sadly, the habitat in which they lived is nearly destroyed and the apes can no longer be studied. Intriguingly, chimpanzees, bonobos, gorillas, and orangutans are proficient tool makers and tool users in captivity.

This mixed picture yields no certainty about whether the common ancestor of African apes and hominids, or indeed the earliest human ancestors, used an impressive tool kit, used no tools at all, or lived in some way that fell between these two extremes. Anthropologists do suspect that hominids made and used tools before 2.6 million years ago, however.[20] Hominids like Lucy, living more than 3 million years ago at Hadar, Ethiopia—very close to Gona and Bouri—had brains comparable in size to those of today's chimpanzees. Judging from the bones, their hands were capable of more sophisticated manipulation than those of modern apes. Grass, twigs, wood, leaves, and other organic materials would have been readily available to fashion into simple tools. It's almost hard to imagine, given what we know of ape cognition and resourcefulness, that Lucy and her hominid kind did *not* make and use tools as they probed for insects or tubers, groomed themselves, and so on.

Excavating a chimpanzee nut-cracking site in West Africa, the archaeologist Julio Mercader and his team found sharp-edged flakes. This was a surprise, since nut-cracking chimpanzees do not set out to create flakes; the flakes were produced as byproducts of the nut-opening process. Because tools occurred at low density, excavators believe that early hominids may have also produced sharp-edged flakes at work sites that are invisible to us now. Lucy's species may well have built on abilities inherited from the last common ancestor of chimpanzees and hominids, and knapped stone in simpler, more experimental ways than was later done at Gona.

We cannot yet ascertain, then, whether *habilis* or some other early hominid invented stone tools. Only with *habilis*, though, did routine making and using of stone tools begin. This fits with the jump in size that characterizes *habilis* brains as compared with all preceding hominid brains. Archaeologists find Oldowan tools in direct association with *habilis* bones *and* animal bones, a situation unprecedented in human evolution. An impressive example comes from Olduvai Gorge. Hundreds of bones are clustered from a wide variety of species—everything from pigs to hippos, from rhinos to horses to elephants. Happily for scientists, markings on the bones show beyond doubt that they were processed by Oldowan tools—and not by carnivores' teeth.

Habilis's increased brain size powered this shift in behavior. Archaeologists think these hominids carried tools to carcass-processing sites, or maybe carried animal bones to places where tools were cached. Either way, this is pretty remarkable behavior: well before the advent of big-game hunting, hominids were able to plan ahead, strategize, and remember past events. This was no run-of-the-mill scavenger; in the ways they processed the bones from already dead animals, *habilis* took a big leap forward in evolutionary terms.

In *habilis* country, stone tools have been found apart from the sources of rock used to create them. From this, psychologists Mark R. Leary and Nicole Buttermore conclude that *habilis* "possessed a more developed ability to think about what it might need to have in its possession in the future."[21] "More developed" here means, compared with chimpanzees. Whether this contrast is accurate is hard to know, but as a hypothesis it makes sense.

In the 1980s, when I was in graduate school, a compelling and rather romantic scenario was all the rage in anthropology. *Homo habilis*, the theory went, not only made and used stone tools for butchering carcasses, but also exhibited social behavior that would be recognizable to us today. During the day, males went off to hunt, and females (along with young children) gathered nuts and plant materials. Every night, the sexes reconvened at a common living area referred to as the home base. There, they exchanged the fruits of their labors. By at least 2 million years ago, then, the human behavioral platform of division of labor and food-sharing was in place; the way the prehistoric stones and bones were arranged on the landscape tell us so.

Before too long, a mighty backlash occurred in response to this theory. Could so much really be read into the accumulated spatter of bones and tools at *habilis* sites? As you may have guessed already, the home-base scenario was based upon an analogy with what living hunter-gatherer peoples do. Unsurprisingly, under scrutiny the archaeological evidence from *habilis* cannot support the analogy. Archaeologists now agree that no good evidence exists for hunting *or* division of labor *or* food-sharing *or* home base at 2 million years ago. Animals were butchered in clever ways, yes, but how did the smart scavengers live? How did they express their social and emotional tendencies? We simply cannot say.

Naturally, this makes it hard to answer our initial question: was the first stone-tool-making event an *emotional* as well as a *technological* one? Admittedly, the term "event" may be a little misleading because, as we have seen, there was probably no eureka moment of discovery. Early hominids in Ethiopia and elsewhere in Africa must have engaged in much trial-and-error experimentation as they mastered the skills of stone-knapping. Nonetheless, somewhere, at some time, an animal carcass was successfully cut open for the very first time by a hominid wielding an Oldowan tool; other "first times" must have occurred in other hominid groups as well. Did the tool makers feel any sense of pride in their accomplishment? Did youngsters try to copy the tool makers' actions, becoming frustrated when the task turned out to be harder than it appeared? Did their frustration turn to celebration when, at last, flakes were successfully struck off the core and used to help procure food?

Paleoanthropologists tend not to ask questions of this type, which is curious because on other topics, speculation runs rampant. As I write, I have just returned from New York's American Museum of Natural History. Though I was there to tour an exhibit on Native American spirituality, I couldn't resist looking into the hall devoted to human evolution. As I had done so many times before, I gazed upon the famous australopithecine diorama, set in Laetoli, Tanzania.

Archaeological analysis of fossilized footprints from Laetoli tells us that two hominids, one larger than the other, walked side by side across an open plain at 3.7 million years ago. The footprints, made by upright walkers, have aided scientists who study the evolution of bipedalism. But in constructing this diorama, the museum chose to put flesh on the bones, or in this case on the footprints, in a provocative way: a male and female hominid walk side by side, the arm of the taller male draped around the female's shoulders.

That's it, you might be thinking? What's so provocative about that? Doesn't it fit the evidence just fine? A confirmed diorama skeptic, I have always thought that the male's posture, together with the somewhat vacuous expression on the female's face, adds up to gender stereotyping: bigger and smarter male protects smaller, vulnerable female. This reaction may be explained in part by my hyperawareness of the historical tendency to highlight the male's role in evolution to the exclusion of the female's, and it may be unfair. But even leaving aside this issue, a problem remains. Scientists have no idea whether the social unit "one bigger plus one smaller hominid" should be translated as "male and female" at all. Perhaps the bipeds were one large and one smaller male, or one larger and one smaller female, or an adult and an adolescent. Why decide to portray a pair bond between an adult male and an adult female?

Other exhibit cases in this same hall reflect the pervasive tendency to tell the story of human evolution in terms of material accomplish-

ments. On view are tool-wielding hominids, hide-processing hominids, and hominids emerging from structures built of mammoth bone.
Twenty years ago, Misia Landau pointed out that human evolution is
most frequently presented as a hero narrative. Over and over, an early
hominid is cast in the hero role and busily overcomes some major ob-
stacle, whether anatomical or cognitive: efficient locomotion (bipedalism); need for technology; how to tame fire; how to bring down big
game; and on and on, in a progressive march toward the modern. Emotional expression is nowhere to be found in this picture.

One of the few archaeologists to take on the problem of "reading"
our emotional past is Sarah Tarlow, who tackles the prehistory of emotions with an admirable mix of innovation and caution. Her article
"Emotion in Archaeology" opened up my own thoughts on this issue.
Projecting our own emotions into the past is easy enough to do, but ultimately unsatisfying and highly unlikely to be accurate. Anthropologists know that emotions are "socially constructed." That is, they do not
exist in some predetermined state within each one of us, waiting to be
unleashed, but rather, despite being rooted in physiology, are formed as
social creatures interact; and emotional expression varies greatly according to the time and place one lives. Because of this, Tarlow writes,
"placing romantic love or late-20th-century styles of anger, jealousy, or
pride in the early 14th century (as in the film *Braveheart*) or deep prehistory (as in the novels of Jean Auel) is hopelessly anachronistic."[22]

Watching fictionalized Scottish history in *Braveheart* and reading
about made-up Neandertals and *Homo sapiens* in Auel's *Clan of the Cave
Bear* and its sequels can be enjoyable pastimes. But Tarlow's point
stands, and it bears real significance for what I am trying to do here. It
would be folly to assert that when Lucy or any other australopithecine
girl (or boy) was born, her mother cradled the baby in her arms and felt
what I felt when my daughter, Sarah, was born in 1993, or what
modern-day Chinese, Fijian, or Ethiopian mothers feel when their
children are newborns. The suggestion that Gona tool makers might
have felt pride in their work may well be a fanciful one, bound to my

own culture and time period. Being proud of something we have made, to say nothing of being *visibly* or *expressively* proud, may not be a human universal today and it may not have characterized our past.

Perhaps in each case—caring for a baby, creating a product—the emotion felt by our ancestors was related more to duty and obligation than to love or pride. Does our inability to choose between these alternatives, or indeed even to come up with the right series of alternatives to consider, doom the whole enterprise? Can we still recognize the commonality of primate emotional experience across the eons? Can we model shifting emotional capacities over time, while refusing to project *specific* emotions back into the past?

One place to start involves the union of the emotional with the cognitive. There's no good reason to wall off the emotional from other aspects of life, even though this has commonly been done since the time of Descartes, who insisted on strictly separating reason and emotion.[23] To grasp what I am saying in modern terms, think about it this way (again, adapted from Tarlow): the sensual can be cerebral (perhaps you feel deep pleasure in solving a difficult puzzle); the intellectual can be embodied (perhaps your logically constructed argument "feels wrong," and you correct it). The brain responds to emotion, and emotion helps sculpt the brain.[24]

The human lineage's early prehistory was steeped in hominid emotion; it had to be. Our ancestors survived in challenging conditions, and not only because they could walk upright, elude a predator, or make an Oldowan tool. They *felt* their lives.

FROM THE MATERIAL TO THE INEFFABLE

Seekers of the earliest traces of religion might be excused for having attacks of jealousy as they review the evidence for evolution of technology presented in this chapter. If only such a sequence as can be traced for stone tools were available for religious artifacts! How satisfying it would be to pinpoint the first highly visible, religious-related behavior at 2.6 million years ago, then trace changes step by step until the appearance of modern human groups. Alas, the search for the religious imagination in human prehistory has no such anchor in the archaeological record.

So what can we do to read religion in the archaeological record, without reading it *into* that record? A vital source for information about the prehistory of religion is Brian Hayden's *Shamans, Sorcerers, and Saints*. Noting that hominids were by definition foragers and not farmers, Hayden reviews the nature of religious ecstasy, ceremony, and ritual in hunter-gatherer people who live (or recently lived) traditionally. He draws conclusions about certain aspects of hominid religion, during certain time periods, based on analogies with these living foragers. When he discusses "defleshing" cut marks made on a skull of *Homo erectus*, hominids who lived after the Gona tool makers (see Chapter 4), he goes a little further than he did in assessing the Makapansgat cobble: the cut marks are potentially the first clear indication that we have of a concept of soul or afterlife.[25] Perhaps as early as 900,000 years ago, he writes, red ocher starts turning up in the fossil record, and probably ancient red ocher "symbolized some sort of supernatural connotations at the broadest level of meaning," as it does for hunter-gatherers today.[26] Hayden theorizes most confidently about religion when he discusses the later hominids, such as the famous Neandertals (whom I discuss in chapter 4). This means he recognizes the limits of the hunter-gatherer analogy for the earliest time periods of human ancestry, accepting its validity only for the later periods.

Yet Hayden's is by no means a consensus approach. In an article entitled "Beginnings of Religion," Ina Wunn concludes that for the time periods of Lucy, *Homo erectus*, and Neandertals, no evidence exists for any religious practice: "All such notions are either products of a certain mental climate at the time of the discovery of the fossils, or of ideologies." Even notions of Neandertal rituals, for example, "belong to the realm of legend."[27]

Hayden's and Wunn's conclusions cannot be reconciled. Wunn's article was published three years before Hayden's book, but the differences between them cannot be attributed to the discovery of new evidence during that interval. Rather, the two scholars interpret the same evidence quite differently. And here we return, yet again, to the realm of emotion. Unlike scholars who study the origins of technology, those who seek to understand the evolution of the religious imagination do not have the luxury of ignoring emotion. Hayden understands this better than Wunn does. As we saw in the first chapter, religion *is* emotion, because it is so grounded in belongingness; it is about feeling deeply for another creature, caring enough for someone to act compassionately, being in awe of the ineffable. To turn away from the emotional life of hominids while trying to explain the development of their religion renders any explanation sterile right from the start.

Is it possible to get at the emotional basis for religion in prehistory? As Hayden beautifully describes in his book, the traditional religions of hunter-gatherer people are deeply experiental, not intellectual:"The main goal of the traditional religious person is to be in contact with sacred forces all the time during this life and to celebrate this connection with ritual."[28] This celebration involves trance, ecstatic states, and shamanic healing—all ways to experience a connection with the universe's vital forces.[29] The evidence for this sort of thing in prehistory will often be indirect, and at certain points the story will be hedged with educated guesswork. At front and center remains a key goal: to remember that these hominids were not only bipedal striders or stone-knappers or cave-wall painters. They were mothers who cared for their children, possibly even speaking to their babies in the singsong cadence ("motherese") we hear in most societies today.[30] They were siblings who romped across the woodlands and savannas of our prehistory; they were meat getters who competed with rival predators to scavenge from carcasses, or to stalk a mammoth, hoping to return to camp not only alive, but bearing protein to share. On the evidence we have from the African apes, it is as likely that they also felt empathy, co-constructed meaning, followed some form of rules, used their imaginations, and were self-aware as it is that they made and used organic tools. Even more likely perhaps; *all* the African apes express these emotionally based behaviors, whereas only chimpanzees, not bonobos or gorillas, make and use tools regularly and in diverse ways in the wild.

If I were able to extend discussion of paleoart from the Makapansgat cobble to manuports associated with *Homo habilis*, this chapter would end on an exciting note indeed. Or if Oldowan tools with intriguing or decorative markings had been unearthed next to *habilis* fossils, debate about early spirituality could really take off. No evidence like this exists, however.

Study of the earliest hominids, ranging from *Sahelanthropus* to *Ardipithecus*, tends to focus on whether they walked upright. In a parallel way, research about early *Homo* is heavily biased toward understanding the whats, wheres, and whys of early stone technology. And despite intense scrutiny, none of the Oldowan tools seems anything more than straightforwardly functional: no decoration or ornamentation of the stone tools is visible. No cores, flakes, or cobbles seem to have been collected for their unusual or striking properties. Yet, as we have seen, early *Homo* is cognitively advanced compared to earlier hominids, and better able to contemplate the future in some basic ways ("If I leave tools here, doing so will pay off later"). Surely this inches us closer to development of a rich imagination, one directed ever further outward from the self. As we will see in the next chapter, as their brains and the surrounding environment began to shift, hominids tapped into that emotional well in new and intriguing ways, ways that involve symbols and that bear directly on the origins of the religious imagination.

FOUR

Cave Stories: Neandertals

R EGOURDOU CAVE, southern France, about 65,000 years ago: a small group of hominids gathers to bury one of their own. Of special importance to the group, the individual who died deserves an elaborate send-off. Chosen members have been at work preparing for the burial ritual.

Now, the time has come. The body, folded into a crouched position, is placed on a series of flat stones at the bottom of a depression. Two leg bones from a bear are placed at the foot of the body, while atop the chest is positioned a slab of rock. Next, a variety of tools is brought to the slab, together with a foreleg bone from a bear, intentionally split in half. The slab, and indeed the body itself, is then covered by a mixture of an ashlike substance with boulders and cobbles. And finally, the entire burial mound is marked with the antler of an elk, and a fire is lit there. After the burial ritual, participants feast on bear meat. As they finish and file away from the grave, some of them already plan their next visit to honor the grave.

For information on Neandertal activity at Regourdou, I am indebted to Brian Hayden's *Shamans, Sorcerers, and Saints*. My French is too spare for the reliable reading of the original archaeological reports that he was able to carry out, and his interview with the site's original excavator, Eugene Bonifay, produced valuable insights. Admittedly, some of my reconstruction here is speculative. Strictly speaking, I cannot know whether group members were chosen to prepare the ritual, or whether bear meat was eaten after (rather than before or during) the burial, or what the hominids involved may have thought or planned at any given moment. Yet in all key respects, this scenario is true to the archaeological analysis and the speculations are highly consistent with it.

Far less famous than its near neighbor Lascaux, where the renowned "Hall of the Bulls" cave paintings are located, Regourdou is a treasure trove of information about this period of our prehistory. Whereas Lascaux was home to early modern humans (see Chapter 5), at Regourdou lived Neandertals, the "cave people" who have piqued our imagination for 150 years. Living in Europe and Asia, so astonishingly like us in some ways and utterly different in others, Neandertals are arguably the most fascinating hominids of all.

To interpret convincingly what happened at Regourdou 65,000 years ago, we must enter the realm of ritual. To begin with: the burial. The placement of the body, its association with tools and stones, and the marking of the grave rule out the work of a natural process. Neither water, nor animals, nor anything related to long-term geological change, could account for these precise arrangements. The dead Neandertal's social companions treated his or her body in a remarkable way, indeed in an unprecedented way.

What role did the deceased Neandertal play in life at Regourdou? Did he or she possess some special knowledge or skills valued by the group? Did participants in the burial mourn their loss, expressing sorrow through tears or gestures or words, or all three? If only we could peer through the millennia and find answers to questions like these, as well as to others that touch directly on the origins of religion. Were the animal bones and tools included in the grave as a way to ease a path into the afterlife? Did Neandertals conceive of some otherworldly, sacred dimension into which they passed upon death?

Resourcefulness helps here, primarily a willingness to navigate by an indirect compass. A focus on the type of animal bones found with the body is a good place to start; understanding what happened at Regourdou is enhanced by knowing about the relationship between Neandertals and bears.

Jean Auel's *Clan of the Cave Bear* may lack scientific rigor when it projects modern human emotions back into our past, but its title is apt. Some Neandertal groups apparently moved beyond a predator/prey relationship with cave bears (a relationship in which Neandertals would have, unfortunately for them, played both roles) into the realm of the symbolic. Consider what is found at Regourdou on the other side of a wall from the burial site: an area called the bear cist.

Essentially a stone coffer, the cist consists of walls and a ceiling. That some of these walls are Neandertal-made is evident because they are vertical and because the stone used to make them came from outside the cave. The ceiling consists of a heavy (more than 1600-pound) limestone slab positioned across the top of two walls. This slab amounts to a final blow to any lingering notion that natural forces were responsible for Regourdou's most intriguing features. "To pretend . . . that this slab could have come to rest naturally in this position," writes Hayden, "over an empty void, without breaking and without crumbling the walls that precariously and precisely support it at its edges, strains credulity."[1] Further, small gaps between the slab and the walls were filled in with smaller rocks, and a second slab partly supported the first. Pierced by a single natural hole, slab number two is distinct from the naturally occurring stone within the cave. Though Neandertals did not create the hole itself, they removed flakes from the slab to make its overall shape more symmetrical.

Inside the cist area are bear bones, carefully arranged: the skull was placed between some stones; long bones were laid out, and shoulder blades were crossed. What may be cut marks from tools appear on the bones, though scientists are not certain about this. The bones are heavily biased toward young bears, as are the bones that show up in caches of animals taken by hunters.

Neandertals took great care, then, in positioning a body and grave goods, in one room, and bones of a bear, in another. What can be learned about Neandertal ritual from these acts? Concluding that the events surrounding the Neandertal burial happened at the same time as those surrounding the arrangement of bear bones in the cist, Hayden describes "a funeral ritual probably involving feasting on bear meat."[2]

The buried body and the bear bones come from bed 4 of the excavation; in other layers are equally intriguing findings that support Hayden's invoking of ritual. Pit V(a), for instance, contains an upside-down bear skull with two cobbles around it, with a third cobble on top; two bear arm bones that were made to cross each other and to lie partly on a rock; and a limestone slab with a hole in it. Once again, the actions of Neandertals are implicated here. And Regourdou is not unique; other Neandertal sites reinforce the picture of intentional burial and symbolic rituals, though in almost all cases the evidence is keenly debated.

Could Hayden be going too far in suggesting a funeral ritual? I was skeptical, after all, of his linking the 3-million-year-old Makapansgat cobble with any hint of supernatural tendencies in early hominids (see Chapter 3). His ideas about Regourdou, however, stand on firmer ground.

First, both the burial area and the bear cist are signposts pointing to symbolic behavior: the slab atop the body is more than just material rock, and the crossed bones in the bear cist are more than just skeletal remains. They represent something—or, rather, the act of positioning them in a special way represents something. It is impossible to understand the archaeology of Regourdou without thinking symbolically, and realizing that the Neandertals were thinking symbolically.

But what of a skeptic's response that the Regourdou Neandertals buried their companion only for some practical reason, maybe to prevent predators or disease from harming their group? Certainly, the Neandertals' large brains could have led them to appreciate, for the first time in human evolution, that there's a cause-and-effect relationship between decaying bodies and approach of predators or the onset of disease. But if the whole explanation is a utilitarian one, why would the Neandertals have placed a slab over the body and an elk antler atop the grave? Why were no other members of the social group similarly interred? Further, if the bear cist contained only the remains of a good meal, the bones, gnawed by hominid teeth, would have ended up mixed up together in a haphazard heap. Instead, they are carefully arranged.

The question then becomes whether it is justified to build on an interpretation steeped in *symbolism* to one based on *religious ritual.* Hayden does link Regourdou to the stirrings of the religious imagination, an unsurprising move given his perspective on prehistory, which we have already encountered. To be fair, he does admit that respect for the dead might by itself explain the Neandertal burial. The burial might amount to a form of ancestor worship; the grave markers would then function to ensure that the group could easily locate the grave and visit it months or years hence. In the end, though, Hayden comes down on the side of Neandertal spirituality, deciding that Neandertals had some notion of a soul, an afterlife, and supernatural power.[3]

Hayden is by no means alone. Year after year, anthropology students in American colleges watch a documentary on human evolution in which Don Johanson terms Neandertal burial "a powerfully spiritual act." In their beautiful book *Prayer: A History,* Philip and Carol Zaleski say that Neandertals enjoyed "a rich prayer life" aimed at securing "the well-being of the dead."[4] The symbolism inherent in Neandertal burials has captured the imagination of anthropologists, religious scholars, and laypersons alike.

Untangling the alternative interpretations of Neandertal symbolic behavior is a challenge to anthropology. How can the various options be cleanly distinguished? Let's review them. First is the respect hypothesis. It may be that Neandertals simply had heightened respect, or a feeling akin to what we call respect, for some individual group members. Perhaps the person was a skilled stone-knapper or plant gatherer as well as being empathetic and thus engendering the loyalty or good feeling of others. The death of such an individual in a group steeped in belongingness would surely have been felt keenly, and commemorated in special, and symbolic, ways.

How odd it seems to discuss Neandertals in these emotional terms! Popular culture portrays Neandertals as shambling, bumble-headed creatures. During U.S. election years, there's no shortage of letters to the editor in national and local newspapers whose writers liken some candidate or other to the dimwitted Neandertal. Neandertal "cavemen" are depicted by the media as club-carrying brutes, sporting animal skins and entirely vacant facial expressions. Even when scholars and museum curators deal with Neandertals, focus is put on skeletal remains or tool making and hunting.

But recall some chimpanzee alpha males are benevolent, and others vengeful; African ape mothers are, typically, highly nurturing of their infants, so much so that infants may pine and grieve when the mothers die. Bipedal australopithecine mothers became acutely sensitized to their infants' distress signals, especially when mother and baby were separated, and tended to them. By the time of the big-brained Neandertals, how much more developed must the emotions have become?[5] Neandertals' emotional connections to each other can never be traced as precisely as can their tools or hunts, but everything we know about primates and prehistory lends credence to the claim that these connections did exist. One of the few archaeologists to award an emotional life to these hominids, Steven Mithen, consigns them to an existence devoid of symbolism![6] What an irony—it seems that Neandertals are *either* emotional *or* cognitively sophisticated in the anthropological literature.

Next on the short list is the ancestor-worship hypothesis. Here, the buried Neandertal is not merely honored at the time of death, but may be venerated for years to come, on into subsequent generations. It's not beyond possibility that the dead one was a shaman or something like it, with a role connected to the supernatural realm. But the evidence for shamanism at this time is thin at best, and for this hypothesis, no such link to the spiritual is required.

What if Regourdou Neandertals did commemorate death in the context of a growing belief in an afterlife? The grave goods, in that case, might be more than markers of status; they might be intended for use by the deceased in a new life, just as in ancient Egypt, kings were entombed with a wide range of items, from furniture to jewels to preserved meats and wine, to ease their new existence in the afterworld. If that was the case, the grave goods might tend to be utilitarian in nature.

Going further, perhaps Neandertals elevated the bear not just to a symbolic status but also to a *sacred* symbolic status. Indeed, it's possible to be endlessly inventive here, reaching deep into the anthropological archives to construct analogies with modern peoples and their animal totems. Until about the 1920s, the Ainu, Japan's aboriginal people, practiced an elaborate bear cult. For the Ainu, the universe was a vital place, teeming with living creatures, all of which possessed souls. Bears were the most important animals of all. An Ainu man would capture a young bear, then feed and raise it in his home; a lactating woman might agree to serve as cross-species wet nurse. After about two years, the bear was ceremonially killed, releasing its soul for rebirth and kicking off a long elaborate feast (on the meat) that would bring together several tribes. Could events at Regourdou have been a very early precursor to such a ceremony?

These alternatives represent, in the order listed, claims for increasingly sophisticated, and increasingly spiritual, Neandertal symbolic behavior. Of course, more than one of these ideas may be correct; expression of a religious impulse in no way forecloses ancestor worship or respect for the dead. While in theory ways to distinguish between these hypotheses could be devised, at present no reliable way exists to do so *in practice*.

Yet now is a good time to revisit a central argument of Chapter 1: *Religion is practice*, based on *emotion in action*. Even if the precise meaning of the symbolic practices at Regourdou cannot be elucidated, we can affirm that they were almost certainly performed with emotion. Respect for the dead, ancestor worship, and involvement with the afterlife share an intersection of the symbolic with the emotional. Whether any of those practices qualifies as religious ritual depends, of course, on one's definition.

For the anthropologist Roy Rappaport, ritual is *the* social act that is basic to humanity. Ritual, he writes, does not just contact the sacred; it creates the sacred. In this way, ritual transforms human action. A piece of cloth that is sacred in one group is not in the least holy in another: the sacredness emerges from the ritual. Operationalizing, Rappaport says that ritual is "the performance of more or less invariant sequences of formal acts and utterances not entirely encoded by the performers."[7]

In some ways this definition seems a bit lifeless; it omits mention of symbols or emotion, a decision seemingly at odds with Rappaport's very thesis about the emotionally intense nature of ritual. But let's look more closely. In his last clause, "not entirely encoded by the performers," Rappaport points to the fact that ritual is repeated over and over, across time. Its essence, its nature, inheres in the fact that a group of people repeats what other people have done in the past. The act honors the past, in a sense. Even when elements change over time, a basic core is carried forward.

As every basic text in anthropology demonstrates, emotion has been central historically to scholars who theorize about ritual; emotion is *the* element that separates the sacred from the merely routine. Of course, Neandertal behavior may not live up to an elaborate definition for ritual, because there is no way to discover what a "formal act" done in an "invariant way" might be, prehistorically. I am comfortable, though, talking about ritual at Regourdou in the same way that I am comfortable talking about the culture of chimpanzees or the technology of *Homo habilis*. There need be no requirement that the definition of Neandertal ritual be precisely identical to the definition of modern human ritual.

But is Regourdou some sort of uniquely complicated site, an outpost of Neandertal symbolism? Before probing this question, let's return to a chronological account of human evolution and fill in some gaps. We pick up where the previous chapter ended: as *Homo habilis* evolved into a new type of hominid.

OUT OF AFRICA

For 5 million years, the great narrative of human evolution was entirely African. Not only did the earliest hominids evolve on the African continent, and the australopithecines and early *Homo* emerge there, the birth of stone-tool technology occurred there as well.

Reflecting upon humanity's deep roots on this single continent brings to mind the feeling I had when walking with baboons on the savanna-grasslands of Kenya. Day after day, I felt strongly connected to the animal and plant life around me. Because of that experience, I understand the desire of some scholars to uncover an innate human preference for the savanna. When shown a range of photographs depicting a variety of landscape types, young children strongly prefer open savanna with trees: this suggests, some argue, that humans, as a species, feel a deep affinity for the landscape where we originated, via a sort of memory written into our genes. I could devote a whole chapter to a critique of this line of thinking. For a start, human evolution unfolded in no single habitat; forested areas were as critical as savanna ones. But the very fact that this research was conducted in the first place showcases the fascination that our African origins hold for us today.

Departing our homeland was a revolutionary step, one taken by neither australopithecines nor *Homo habilis*. *Habilis*, the first hominid in our genus, evolved from an earlier gracile form of australopithecine at about 2.4 million years ago, then endured for about half a million years. As we have seen, *habilis* strategized about how to bring together animal bones with Oldowan tools, an ability that made their foraging easier and more efficient. That the first *Homo* brains were larger than those of any early hominid (and all living apes) was a key factor in this cognitive shift.

At 1.8 million years ago, the pace of behavioral evolution began to quicken. No innovation is more intriguing to our highly mobile

twenty-first-century society than long-distance migration. Perhaps
driven by the need to follow game, or in search of other resources,
hominids began to leave Africa.

Who were these wanderers? Since their discovery on the island of
Java in the 1890s, they have been called *Homo erectus*. More than a quar-
ter century before Raymond Dart held a tiny skull in his hands and
proclaimed the antiquity of bipedalism, *Homo erectus* fossils were
thought to represent the very first bipedal ancestor. The island of
Java continues to play a pivotal role in scientific understanding of this
hominid, but *erectus* was distributed widely, from Olduvai Gorge to
caves near Beijing.

Some anthropologists say that the African form of *erectus* differs
enough from the Asian one to be credited with its own species name.
A split taxonomy results from this scheme, with *Homo ergaster* in Africa
and *Homo erectus* in Asia. A debate of this sort is critical for taxono-
mists and paleoanthropologists, but not for us: a single term suits our
purposes just fine. The key point is that at about 1.8 million years ago,
a new suite of adaptive characteristics emerged.

Some trends that were first evident in *habilis* now continue in *erectus*. Brain size increases markedly, so that some *erectus* skulls are as large
as those of some people living in contemporary times. I suspect the
writer Anatole France would rather be remembered for his Nobel
Prize-winning literature than for his brain size; to be sure, his satire
Penguin Island should intrigue students of evolution and of religion
alike. Yet there's no getting around the fact that France's brain—
measured at 933 cubic centimeters after death—was unusually small,
and as such, useful to illustrate a point about *erectus*.

Were you able to measure the brains of your family and friends,
the average result would probably come in at about 1,350 cc, or about
three times the size of the brains of today's chimpanzees and gorillas.
Yet the range of modern brain size is vast, and smaller-brained folks are
no less intelligent than anyone else. No slouch in the intelligence and
imagination department, Anatole France, with his small brain, falls at

the lowest end of the modern continuum. (Brains may shrink as people age, but scientists estimate that France's brain, even when he was a young man, could not have much exceeded 1,000 cc.) The point to emphasize is that France's brain was no match, size-wise, for the brains of some *erectus* individuals.

There's no sense, then, in trying to predict cognitive potential by measuring brains. Nonetheless, in broad evolutionary terms, cognitive leaps often do follow brain-size increases, and it's no different with *erectus*. As before, stone-tool technology is the most concrete evidence. A teardrop-shaped implement called the hand ax is the quintessential *erectus* tool. Easily distinguished from an Oldowan cobble or flake tool, the hand ax has flakes chipped off all sides. In its form, the hand ax reveals both the sophistication and the limitation of *erectus* intelligence.

The most arresting feature of the hand ax is visible once specimens from across Africa are lined up and compared. The comparison is limited to Africa because no hand axes have been found with *erectus* in Asia. This state of affairs led to an enduring supposition that Asia was a technological backwater in human evolution. This assumption died a swift death when tools whose workmanship was comparable to that of the African *erectus* tool kit—but minus the hand ax itself—were found in south China and dated to 800,000 years ago. These so-called Bose Basin tools exemplify a link between environmental change and behavioral innovation by *erectus*. Right around 800,000 years ago, a meteorite struck this part of Asia, setting into motion a domino-like set of events: the meteorite caused fires, the fires cleared out tree and bush cover, and a new source of stone was exposed.[8] Before long, *erectus* was fashioning this stone into new tool types.

When tracing events in human evolution, it's almost impossible not to be diverted down these ancestral cul-de-sacs. To return to the African hand axes: careful scrutiny reveals that their form is standardized, as the forms of Oldowan tools were not. Hand-ax makers chose, over and over, to create a *particular* shape. From this observation, a bloom of questions follows: Was the preferred form taught by elder to

apprentice? Did teacher and student discuss how to go about the craft-
ing process using words, gestures, or both? How did the form become
standardized across groups and over long distances?

Yet, in the degree to which this standardization persists over time
as well as through space, the hand axes' limitations are revealed. For
more than a million years, a hand ax was a hand ax was a hand ax; no
improvements were made whatsoever. *Homo erectus* fashioned cleavers
and picks suited to other jobs, and thus invented a true tool kit. Perhaps the existence of such a toolbox relieved any pressure to tweak the
hand ax, which might have worked just fine as it was for a certain set
of tasks. Still, it's hard to imagine a comparable level of stagnation for
many implements in the current age, when a four-year-old computer
is an outdated computer. How likely is it that even the historically
durable hoe or shovel will persist in its current form beyond the
twenty-first century? In any case, the astonishing pace of technological
innovation today has no roots in the tools of our *erectus* ancestors.

This mix of advances and limitations is a theme with *Homo erectus*.
For decades, debates have raged about whether *erectus* made fire and
hunted big game. The best consensus at present is that limited fire use
was in place but that organized big-game hunting came only later. That
last sentence is packed with qualifiers. It's meant to leave open the possibility that *erectus* controlled fire in some places and engaged in small-
scale opportunistic hunting at some times. Yet fire making and hunting
are highly unlikely to have been fundamental to the successful *erectus*
adaptation.

A sampling of sites on three continents gives a feel for the nature
of this adaptation. About 1.6 million years ago in Kenya, a twelve-year-
old boy died. The cause of death may have been an acute infection,
judging from the oral abcess evident in his remains. Upon their discovery in 1984, the boy's bones were catapulted into fame because at the
time they represented the most complete hominid ever found—twice
as complete as Lucy. Other *erectus* individuals as much as 1.9 million
years old have been found in East Africa, but none gives to paleoan-

thropologists the joy in "reading the bones" that this boy does. What do his bones reveal?

The so-called Nariokotome Boy would have topped out at around six feet in height, had he lived to adulthood. From the neck down, his skeleton is quite robust but basically modern, especially in the limbs and the pelvis. From the neck up, the look is far more primitive: his jaw is heavy, his brow ridges protrude, and his skull is shaped quite differently from ours, with a low forehead. To a trained anatomist, it is clear that the Nariokotome Boy's overall build was ideally suited for life in the tropics—long and lean, like the bodies of modern Masai people from equatorial Africa. Their bodies evolved to dissipate heat, whereas the shorter, chunkier Inuit or "Eskimo" peoples evolved to conserve it. The Nariokotome Boy's bones also tell us that *erectus* was committed to walking on the ground, rather than to switching flexibly between ground-walking and tree-climbing as had earlier hominids. Though no stone tools were found with the boy, hand axes were in evidence in *erectus* African populations by this point.

Given all these indicators of successful adaptation, it might seem as if *erectus* should have been perfectly content to stay put in Africa. Yet by the time of the Nariokotome Boy's life and death, some *erectus* individuals had already left Africa and settled in Europe and Asia.

In the ex Soviet republic of Georgia, there's a town called Dman isi, about an hour's drive from the capital of Tbilisi. A key stop along the Silk Road at one time, Dmanisi is now a pleasant diversion for tourists interested in medieval history. And the locale was apparently attractive to our ancient ancestors, as well. In prehistory, animals of all sorts and sizes roamed over a lush savanna here. *Erectus* turns up at 1.8 million years ago, barely after its first appearance in Africa as a distinct species.

Dmanisi, together with a handful of other sites, turns old theories inside out. Originally, scientists thought that *erectus* incubated in Africa for hundreds of thousands of years, later traveling north to Asia and, much later, on to Europe. Now, we know otherwise. In Europe, evidence at Dmanisi is clear. In Asia, at Mojokerto on the island of Java, *erectus* was present at about the same time. There's even evidence in central China's Longgupo Cave that *erectus* may have reached there as early as 1.9 million years ago. Was something pushing certain populations of *erectus* out of Africa? Were they propelled in their migrations by technology?

The Dmanisi hominids have something to say on this score, for their tools are of the Oldwan type, that is, core-flake tools more typically associated with *Homo habilis*. No hand axes have been unearthed at this site. Apparently, as useful as hand axes must have been for our ancestors, they were not required for intercontinental travel. If *erectus* migrated because they were following game animals, it is highly unlikely that they did so as serious, organized hunters.

Mastery of fire, too, would have been an obvious boon to the *erectus* travelers, but here again, the early dates jeopardize any such link. Later, by about 1.5 million years ago, hominids living in South Africa were probably controlling fire. At the site of Swartkrans can be found the earliest convincing evidence. Natural fire burns at lower temperatures than does human-set (or hominid-set) fire; at Swartkrans, the fire was so hot it is almost certainly hominid-set. Australopithecines as well as *erectus* lived at Swartkrans. If *erectus* invented fire here, it was well after the initial migrations to Europe and Asia.

A single skull found at Mojokerto, Java, excites scientists for what it reveals about brain development. It comes from an infant between six months and eighteen months old. Using CT scans, scientists have determined that the growth rate of this infant's brain paralleled that of apes and australopithecine brains. That is, the rate was quite fast. In *erectus*, the intense period of brain growth takes place before birth, as is typical of most mammals—but not of humans, where the first year after birth is characterized by major brain growth in the extra-uterine "incubator" of parental attention and enrichment. Given that *erectus* is well adapted to its own time period and niche, that its biology differs from human biology should not be surprising. Does this apelike growth pattern signify that *erectus* infants were markedly less dependent on their caregivers for learning about the world than *Homo sapiens* infants were, and are? The Mojokerto infant skull is only a single case, from early in the *erectus* time span, and thus cannot answer a question such as this.

Scientists are certain—at least, as certain as is possible—about the 1.8 (or maybe 1.9) million-year-ago date for the origin of *erectus*, but the species' end point is shrouded in uncertainty. Some experts say that *erectus* lived until about 30,000 years ago, whereas others think that *erectus* developed into new species much earlier than that.

To add to the confusion, other hominids lived during the reign of *erectus*. In northern Spain, a spectacular site called Gran Dolina points to the presence of some type of *Homo* successfully butchering deer, bison, and rhino at 800,000 years ago. Most experts agree this was not *erectus* but a different type of hominid (*Homo antecessor*), which suggests that variation within the *Homo* lineage was pronounced. (The robust australopithecines overlapped with *erectus* too, but the robust lineage and the *erectus* lineage were visibly distinct, so scientists would not confuse a bone of a robust australopithecine with an *erectus* bone.) In sum, it can be quite difficult to discern whether a certain fossil should be classified as *erectus*. For now, we can make an end run around this problem by choosing Zhoukoudian Cave in China as our final site, because it is definitively linked to *erectus*.

Zhoukoudian has been a center of intrigue in anthropology for over sixty years, and our understanding of this site has shifted a great deal during that span. Even its name has changed. When, in 1979, the Chinese government adopted a new system for rendering Chinese characters into Roman letters, Peking became Beijing, and Choukoutien became Zhoukoudian. For anyone trying to grasp our behavioral past, Zhoukoudian is seductive, for it brings together hominid bones, animal bones, and evidence of fire. Some of the bones are charred. Analysis of the cave shows that the *erectus* bones span a long period, from about 670,000 to 400,000 years ago.

Generations of students were taught that the cave was a home base for *erectus*, indeed for generations of *erectus*. Recall our discussion of *habilis* and you'll get the idea: Zhoukoudian was thought to be a haven from predators, a living area where children could be raised safely. The *erectus* cave-dwellers were, in this scenario, masters of fire; after hunting down prey, they cooked the meat and consumed it safely in their shelter. It was even asked whether the burned bones might hint at a much darker side to these hominids: were *erectus* eating each other, in some kind of cannibalistic ritual?

Much of this imaginative thinking about *erectus* has faded from anthropology texts, like the romantic scenarios of *Homo habilis* as home-base-dwelling food sharers. It seems that Zhoukoudian Cave was just as likely to have been a hyena den as a haven for *erectus*, and that the hominid bones were carried in by hungry jaws. Some of the fires that burned in the cave were probably spontaneous, not set by *erectus*. Even the presence of charring on the *erectus* bones is now in dispute.

Not all the evidence at Zhoukoudian is contested, however. Burned animal bones are found together with hominid-made stone tools. On the bones are tool marks that *overlie* any gnaw marks made by carnivore teeth, and that therefore suggest hominids were able to scavenge meat from carcasses. If Zhoukoudian hominids were not the mighty hunters of earlier imagination, they probably did roast meat, scavenged from carcasses brought down by carnivores.

Leaving China, let's once again take a broader view, and assemble an overall picture of *Homo erectus*. Unless *habilis* or australopithecines are someday found outside Africa, *erectus* will always carry the banner "first hominid traveler." Looking at the evidence one way, we can see that over time, *erectus* engaged in behavior related to fire and carcass-processing that implies increasingly cohesive social groups. It's easy to picture a group at Zhoukoudian, for instance, setting up a division of labor to accomplish a variety of tasks. Some *erectus* would have been the tool makers; others, responsible for disarticulating a horse or deer carcass; still others, for starting a fire and doing the cooking. Perhaps guards were stationed at the cave entrance, to monitor the approach of hyenas. Perhaps children were apprenticed for short periods in any or all of these tasks.

And what about emotions? Imagine the earliest *erectus* on the move. Pretty clearly, this wasn't a case of hominids in Kenya sitting around, debating, then deciding to move to Europe or Asia. Travel was no doubt slow, with one generation making it just so far, the next generation continuing on, and so forth over thousands of years. Perhaps no single group encountered a dramatically new environment during the lifetimes of its members. Still, long distances were involved and new areas were colonized. Could hominids have done all this without feeling, along the way, curiosity, fear, worry, and excitement, and without feeling affection for some traveling companions but dislike for others?

Looking at the evidence another way, we can see hominids with what seems to be fairly rapid apelike brain development, no apparent capacity to improve upon the efficiency of their basic tool (the hand ax), and no burial of their dead. *Erectus* was not routinely a big-game hunter, and did not feast on megafauna carried back to a protected home base. It is always possible, of course, that future discoveries will bring surprises on any of these scores. Four hundred thousand years ago in Germany, hominids created wooden spears about six feet long that were clearly used as hunting weapons. The spears were found at a site called Schoningen, among stone tools and animal bones (including horse bones), and amount to calling cards for some type of long-distance hunting technique. Whether the Schoningen hominids are *erectus* or not is still unclear, but the time period is right; it may turn out that *erectus* began coordinated hunting at around this time in some regions.

For many anthropologists, a haunting question remains: could any of the most intriguing *erectus* behaviors—migration, control of fire, cooking, standardization in hand-ax manufacture—have been accomplished without language? Were these hominids able to express their emotions verbally? The evidence is patchy, yet the most convincing arguments for evolution of language take a continuity approach. If technology, the expression of belongingness through emotional engagement, and indeed, religious expression, all evolved gradually, why not language? I'll have more to say about hominid language later.

RELIGION

Nothing even vaguely akin to the Neandertal funeral at Regourdou is known for *Homo erectus*. Some hints of incipient ritual can be found but not tied definitively to this species, as we can see from the Bodo cranium. Dated to 600,000 years ago, the Bodo skull from Ethiopia is considered *erectus* by some, and a later transitional species by others. It is definitely from a pre-Neandertal species of some sort, and at a minimum, closely related to *erectus*.

The Bodo skull was "defleshed," meaning that its skin was removed intentionally. Just for fun let's peek at a technical report on this skull: "The symmetry of the oblique cut marks on the frontal region and the consistent parasagittal directionality and dual-track morphology of cut marks on the posterior parietals argue for a patterned intentional defleshing of this specimen by a hominid(s) with a stone tool(s)."[9] (Who says too few American scientists are bilingual?) Jargon, translated: the visible marks on the skull are of such a nature as to rule out damage from weathering, animal gnawing, or the action of stones or hooves; they point toward some practice in which flesh was taken off this cranium by design.

What practice might this be? An obvious possibility is some kind of mortuary behavior. Hayden comments that it's hard to think up any practical reason for defleshing, "since there is almost no meat on the skull." Far more likely is a symbolic practice of some sort, perhaps involving "some concept of a spirit essence in the skull. This may be the first clear indication that we have of a concept of soul or afterlife."[10]

Might defleshing be a sign of cannibalism? Here is a fraught topic if ever there was one. Many people have a hard enough time accepting our evolution from a common ancestor with apes, let alone our evolution from a hominid who ate his rivals or even his companions. But, as it turns out, the Bodo skull can tell us nothing on this score. A good sign of cannibalism is when the brain is extracted; the Bodo cranial base is missing and can't speak to brain extraction one way or the other.

My own view is that it cannot speak to the spirituality question, either: just as no clear link exists between the defleshing and cannibalism, so none exists between the defleshing and a sense of spirituality.

Not long after the Bodo defleshing, at around 400,000 years ago, "the world's oldest sculpture" was created by a hominid (or hominids), perhaps *erectus*, living in Morocco. Or at least that's how some have interpreted a small quartzite figure found, close to some tools, on a riverbank near the town of Tan-Tan. The figure resembles a human form, at least to modern Western eyes, but this overall shape is just a happy accident. Hominids do seem to have modified the object, however, in intriguing ways.

Robert Bednarik, whose expertise aided our understanding of the Makapansgat cobble from earlier prehistory, has described eight grooves in the Tan-Tan figure that serve to accentuate its anthromoporphic nature, specifically in the neck, arms, and legs.[11] Based on close analysis and experimental reconstruction, Bednarik believes that hominids intentionally amplified, or emphasized, these grooves. The case for hominid modification of the stone is strengthened by traces of hematite, a red pigment, on part of the figure. (Red again: recall the Makapansgat cobble.) If Bednarik is right, the Tan-Tan figure is an instance of paleoart.

Whether this counts as the "oldest sculpture," as proclaimed in the headlines, is in the eye of the beholder. Skeptics have not been silent. If, in fact, the supposed accentuations were natural, and not made by hominids, what's left is an impressive ability by members of a species of later *Homo* to perceive their own image in the prehistoric natural world. If the Tan-Tan figure is indeed modified, we would want to know whether the modification is an expression of the religious imagination. Was the figure used in some ritual? By now, my answer to this question is predictable. Yes, the figure points to our ancestors' immersion in a world of symbols, a world of representation. And perhaps the Tan-Tan figure was incorporated into some sort of ritual, even a sacred ritual. But a leap from symbol to ritual to *sacred* ritual is again too great to make in good conscience on the basis of a single artifact. As we will see, it's only when evidence is considered collectively that it begins to speak promisingly to the origins of the religious imagination.

NEANDERTALS AND OTHER ARCHAICS

Between the time of Zhoukoudian Cave with its undisputed *Homo erectus* fire makers, and that of the Regourdou site with its Neandertal burials is a somewhat shadowy period of human evolution. The so-called archaic hominids were more humanlike than *erectus* but not as advanced as either Neandertals or modern humans.

Archaeologists have uncovered a "cave of bones" at Atapuerca in northern Spain dated to around 350,000 years ago. At the bottom of a vertical shaft in the cave is a pit jammed with more than two thousand hominid bones representing at least thirty-two individuals. The skeletal material indicates that most of these individuals were young adults who died at about the same time; their bodies were intentionally dropped into the pit. Perhaps the Atapuerca community experienced the massive die-off that accompanies a famine that first kills off the vulnerable young and elderly, and then, finally, these prime-age individuals. Did the individuals who disposed of the bodies participate in some kind of death ritual? We do not know, yet Mithen is surely right to imagine that the survivors felt "intensely emotional" about the loss to their community.[12]

By about 250,000 years ago, true big-game hunting was unmistakably underway, as on the Channel Island of Jersey where hominids hunted woolly rhinoceros and mammoth. The hunters drove their prey over a cliff, then butchered and ate the meat. This period, too, boasts an occasional artlike symbol that forces us to think hard about its meaning. From Israel comes a basalt figure, dated to 230,000 years ago and anthropomorphic in shape. The Berekhat Ram figure, like the one from Tan-Tan, is best understood to be a natural form with some modification by hominids, including the application of red ocher. Indeed, by the time of Berekhat Ram, red ocher is in use by hominids at a number of locations, a practice that Hayden stresses is "strictly decorative and symbolic."[13]

Any glimmers from archaic hominids of symbolic and *potentially* ritual behaviors pale beside what is known about Neandertals. Burials; the decoration of bodies with jewelry and pigments; cannibalism; cults; face-to-face interaction with *Homo sapiens*—Neandertals have it all, and it's little wonder that they thoroughly capture the popular imagination. As key players in the story of human evolution, Neandertals will span this chapter and the next.

Key players they are, but not key ancestors. Anthropologists now pretty much agree that Neandertals are not directly ancestral to *Homo sapiens*, but are an offshoot of the human lineage that ended in extinction. So far, I have chosen to avoid wrangling with the diverse, highly contentious theories about which hominids did and did not lead to modern humans. Yet the pattern is clear: *Homo sapiens* evolved from a version of *Homo erectus* and/or its archaic relatives. *Homo erectus* itself evolved from earlier *Homo*, whose own ancestor was some version of a gracile australopithecine. The robust australopithecines have long been identified as a specialized adaptation, leading only to extinction.

Now DNA speaks to us through the millennia to say that the Neandertals, too, were a dead end. Like the robusts, they contributed nothing to the modern human gene pool. The Neandertals differed from the robusts in one vital respect: they coexisted with modern humans. In the Middle East, they did so for a long period, though whether Neandertals and *Homo sapiens* interacted face-to-face there is unclear. In Europe, the two species overlapped for at least 10,000 years and certainly met each other. Even as nonancestors, then, Neandertals must have shaped our own evolution in significant ways.

What a stretch of our much-vaunted modern mind it requires to imagine sharing the earth with another hominid species. Decisions about where, what, and how to hunt may have been affected by what "the other guy" was doing. Cultural exchange of tools and tool-making techniques likely took place between Neandertals and *Homo sapiens*.

Could this level of interaction have occurred without emotional response on both sides? This part of the Neandertal story is best reserved
for the next chapter; first needed are the basics of anatomy and behavior, and a return to Neandertal ritual.

Given the emphasis in this book on our deep African roots, an
alert reader is by now wondering why I have made no mention of
Neandertals in Africa. Surprisingly enough, Neandertals are found
nowhere on the African continent. Archaic hominids before the time
of modern *Homo* do live there, of course; a fairly unbroken fossil record
exists from late *Homo erectus* until the first *Homo sapiens*. But Neandertals are restricted in their distribution.

Hominids recognized as fully Neandertal lived from about
130,000 to 28,000 years ago. Before that, back to about 300,000 years
ago, hominids showed some elements of the Neandertal adaptation,
but not the entire complex of features. Happily for scientists, this full
complex is so striking as to be readily recognizable, even diagnostic.
Neandertals were robust, stocky, cold-adapted hominids with thick
bones, big brains, and a cluster of anatomical oddities, notably the
shape of the skull and of the brow ridges; the size of the nasal cavity;
and the large and worn front teeth.

"Harsh" is the best single word to characterize Neandertals' lives.
Stress is inscribed on their bones in the form of healed fractures and
the marks of degenerative disease. From the cave of Shanidar in Iraq
comes a famous example, immortalized as the character Creb in Auel's
Clan of the Cave Bear. This individual (called Shanidar I in the anthropological literature) suffered from severe arthritis. Most of his teeth
were gone; one eye orbit was crushed; and his right arm was withered,
and had been for many years. Though Shanidar I was in sorry shape,
none of these assorted ills caused his death.

So how did Shanidar I, and others with similar disabilities, get
along in a harsh world? One view is that he benefited from the empathy and compassionate action of his companions. It's easy enough to
picture this: hunters and gatherers from his group might have shared
food with him on a daily or weekly basis. Perhaps he stayed sheltered

in the cave, performing some domestic duty of use to the group. Or perhaps he was literally supported—aided by others as he walked—during group expeditions.

The anthropologist Katherine Dettwyler has challenged this view by pointing out a sobering fact: sometimes, disabled individuals are barely tolerated by other members of their society; sometimes, they are treated with outright cruelty. Neglected, or worse, they may still manage to eke out a living. It is impossible to know whether Shanidar I enjoyed a cushioned existence or barely managed to survive in the face of cold indifference or cruelty.

Neandertals did not live beyond their mid-forties. They appreciated beauty and symbols, and they participated in ritual, but the overriding fact of their existence is that they lived brief, difficult lives. (It's worth pointing out that in a handful of countries in sub-Saharan Africa today, life expectancy is forty years or less. Sadly, life expectancy is severely constrained when access to the world's resources is limited.)

The archaeological record of Neandertal behavior is rich and fascinating. Neandertals hunted successfully, bringing down even prime-age adult bison and aurochs. They controlled fire, constructing hearths at certain sites. They created tools far more sophisticated than hand axes. Made from a so-called prepared core, these amounted to a series of flakes sliced off, assembly-line-style. This efficient process produced flake tools fashioned specifically for particular tasks at hand. In some places, Neandertals made tools more complex even than this, and virtually indistinguishable from those of *Homo sapiens*.

When it comes to interpreting Neandertal symbolism and ritual, the intellectual rivalry in anthropology ratchets up a notch. *Scientific American* published a special issue a few years ago, taking a "New Look at Human Evolution." Gracing the front cover is an artist's rendering of a very early hominid, covered with fur, but with an intelligent gleam in the eye. Between the covers of this issue can be found a wide range of opinions on the behavior of Neandertals.

First, Ian Tattersall, a curator at the American Museum of Natural History who is also a well-known writer: "Despite misleading early

accounts . . . no substantial evidence has been found for symbolic behaviors among these hominids or for the production of symbolic
objects—certainly not before contact had been made with modern humans."[14] Here is an echo of Steve Mithen's judgment that Neandertals
did not live a life steeped in symbols. An arrestingly different view is
expressed by the European prehistorians João Zilhão and Francesco
d'Errico:"The behavioral barrier that seemed to separate moderns from
Neandertals and gave us the impression of being a unique and particularly gifted human type—the ability to produce symbolic cultures—
has definitely collapsed."[15]

The distance between these two opinions measures what I love
about what I do for a living. Much is up for grabs in the interpretation
of primate and human behavioral evolution, and the fun for anthropologists lies in digesting divergent accounts and deciding for oneself what
to think. In this case, I come down firmly on the side of Zilhão and
d'Errico. Why? Here is a shortlist of Neandertal symbolism, even leaving aside symbols associated with burials:

- At a cave site in northern France, Neandertals living 33,000
  years ago modified teeth of bear, wolf, and deer, and wore them
  as pendants. This site, Grotte du Renne (from an area known
  as Arcy-sur-Cure), is home to Zilhão and d'Errico's research.
  Homo sapiens in Europe at this time wore decorative jewelry,
  too. When *sapiens* created jewelry, they pierced the material
  used in order to suspend and wear it. Neandertals did no
  piercing, but modified the base of the tooth they wished to
  hang, then tied a stringlike material around it.
- Also in France, but a little earlier than these pendant-makers,
  Neandertals modified a triangular piece of flint to bring out its
  facelike qualities. Flakes were removed selectively and a splinter
  of bone was rammed through the hole and secured there by
  two pebbles, making the hole appear eyelike. This object,
  referred to as the French Neandertal mask, continues a theme
  stretching back to the Makapansgat cobble and the Tan-Tan

and Berekhat Ram figures: an apparent fascination with facelike properties in the natural world. This mask has been anointed by Paul Bahn, an expert in prehistoric art, as the item to "finally nail the lie that Neandertals had no art."[16]

- Neandertals created and appreciated music. The evidence of their general level of intelligence, emotional expression, and patterns of cave use makes it highly likely that they sang and danced in caves, clapped together bones and stones in rhythm; and appreciated "the melodies and rhythms of nature" all around them.[17] Suggestions of ancient musical instruments do not withstand scrutiny, however. In Slovenia, archaeologists discovered the leg bone of a bear in a shape that resembles a flute, with four holes perforating its top. The bone was dated to 40,000 years ago. Enchanted by what they were quick to label as the world's oldest known musical instrument, science writers imagined a group of Neandertals listening to sweet tones played by one of their group around an evening's campfire. Microscopic analysis suggests, instead, that the holes were gnawed by carnivore teeth.[18]

- At Tata, Hungary, Neandertals made a plaque that can only be decorative and symbolic. The plaque is a mammoth molar that was polished, smoothed, and covered with red ocher; it is likely over 100,000 years old, though this date is somewhat uncertain.

What does this symbolism mean for our quest to learn something about the emotional nature of Neandertals, and whether that nature extended toward the spiritual? The question isn't aimed at whether Neandertals were composing music, or creating what would pass for art in the Louvre or the Hermitage or even in a kindergarten class in New York or Nairobi. Rather, it acknowledges the evidence that at some sites, Neandertals engaged in wholly nonutilitarian behaviors, and it asks whether these behaviors went beyond the symbolic to the religious.

The picture of bare survival in a harsh climate is not accurate for all Neandertals. These hominids did not devote all of their time and effort to hunting, following game, gathering plants, tool-making, fire-making, and tending to their children. Some of them engaged in what we think of today as creative, even soul-stirring acts . . . a phrase that brings us right back to ritual.

RELIGION

At the heart of some stereotypes lies a grain of truth. To think of Neandertals as "cavemen" is simplistic, not to mention sexist; the mental picture of club-wielding Neandertal males dragging their female mates into a cave is best forgotten. But at some sites, Neandertals did live in caves, or routinely utilized them. Analysis of Neandertal cave life points to elements of a burgeoning religious imagination fed by participation in ritual.

From the cave at Shanidar, Iraq, comes the most famous Neandertal burial of all. A favorite of textbook writers and documentary filmmakers for decades, Shanidar is often nicknamed the "flower burial." Fossilized pollen indicates that twenty-eight different types of flowers, ranging from thistle to grape hyacinth, are represented in the Neandertal burials in this cave. (Nine skeletons were found in the cave; four had been deliberately buried by Neandertals.)

Re-creations of the Shanidar burial often refer to a spiritual act, one in which Neandertals are moved to send their dead into an afterlife with beautiful flowers. Reasonable doubt exists, however, about the source of the flowers, because rodents may have brought the pollen into the cave. Scientists are locked in debate about this point;[19] meanwhile, Shanidar is best listed in the "questionable" category for ritual behavior.

Not all sites are so ambiguous. At La Ferrassie in France, a Neandertal skeleton was covered with a limestone slab bearing hominid-created markings. At Teshik-Tash in Uzbekistan, a Neandertal child was buried encircled by goat horns. But even at these sites a way is

needed out of the deadlock: how can a real connection be forged between ritual and religion?

Caves may still hold a clue. I have never been a spelunker; my only
experience with cave exploration comes from group tours of Luray
Caverns in Virginia, where I live. Descending into the caverns is, by
any standard, a modern and managed experience: electricity lights the
way and footpaths ease the rough ground. Even so, by peering down
forbidden passages into the heavy gloom and imagining away the gar-
rulous tour guide, it is possible to conjure up a sense of the depth and
blackness of prehistory's undisturbed deep caves.

Unlike any hominids before them, Neandertals began to develop a
sense of deep caves as special places, and quite probably as places for
ritual activity. Far inside Bruniquel Cave in France is a curious structure of stalactites and stalagmites. That Neandertals constructed this
arrangement is certain, though what it might be is anyone's guess. It includes a hearth that was lit for fire at about 50,000 years ago. Similarly,
from a cave in Arcy-sur-Cure, the place where Neandertals made pendants from animal teeth, comes evidence of a hominid-made structure
far from the cave's entrance. In an area very narrow and only a meter
high were tools, broken bones, and a mammoth tusk. Surely this dark,
cramped area was not a part of the cave where Neandertals lived or carried out daily tasks.

From the use of these caves, from the act of entering "a timeless
place where senses and consciousness become altered," a mystical sense
connected to a world beyond the here and now may have emerged.[20] I
believe it is overwhelmingly likely that in such a situation Neandertals
would have expressed religious imagination in some form.[21] Indeed,
Neandertals may have been not only the first hominids to make jewelry, appreciate music, and express their emotional commitments to
others through ritual burial of the dead, but also to pray. If we understand prayer as active communication between human and divine
realms, then there's every reason to think that caves may have been
places of prayer for Neandertals.[22]

After all, we are taking the view that religion is about action, and we know that for native peoples and forager peoples even today, the sacred is bound up with daily activities. Hunting is sacred, in and of itself, for these people. As Neandertals coped with harsh climates, large predators, physical trauma, competition from *Homo sapiens*, and the mystery of death, they may well have begun to enter into this relationship with a world beyond the present, a world of forces to which they felt an intimate connection but which they did not fully understand.

Overly romantic views of the past should be avoided, though, as earlier chapters have shown. To push too hard for a species committed to flower-strewn, afterlife-oriented burial rituals would be a big mistake. Any notion of continuous emotional harmony among Neandertals is far too simplistic, as it would be with respect to any sentient creatures. For one thing, whereas the evidence about *erectus* and earlier hominids is ambiguous, some Neandertals clearly ate their own kind, an act not easily squared with romantic reconstructions of the past.

At about 100,000 years ago, Neandertals living at the French cave site of Moula-Guercy on the Rhône River defleshed and disarticulated carcasses of other Neandertals in the same way that they defleshed and disarticulated carcasses of deer.[23] Soft tissues and marrow were removed in each case. In the absence of any indication that this behavior was related to mortuary practices, interpreting it as something other than cannibalism is a real stretch. Indeed, the chief criterion for deducing cannibalism is met: hominid bones were prepared for eating just as were bones of other mammalian prey.

Though the evidence at Moula-Guercy may be the most convincing, it is not unique. In Croatia, for instance, the sites of Krapina and Vindija point to cannibalistic practices among Neandertals. As the paleoanthropologist Tim White puts it, most anthropologists no longer ask whether cannibalism existed, but why.[24]

Few questions strike more closely at the heart of this book. White is impressed by the fact that cannibalism occurs at only certain Neandertal sites, and deliberate burial only at others; he considers this variability to mirror modern behavioral diversity.[25] But a critical question still begs for an answer: was there a link among burials and bear cults (as at Regourdou), deep-cave activies (as at Bruniquel) and cannibalism (as at Moula-Guercy)? All three behaviors did occur in France, but it is a conceptual link I am after, not a geographic one. All three behaviors may express different facets of an early engagement with the sacred.

Many anthropologists think that Neandertals had some sort of language, though possibly a language not much like ours. The Neandertal voice box or larynx seems to have been in a position that would enable speech. (This throat anatomy may even have evolved by the time of later *erectus* populations.) Added to this is the collective nature of Neandertal behavior, whether in big-game hunting or burial rituals, which seems to strengthen a claim for language. Language is social behavior: it serves to bring people together, to transform individual action into collective action. When ritual creates the sacred today, it often does so through language.

We can almost grasp something solid here, something to point definitively to a relationship between Neandertals and sacred aspects of the universe. It is as if a veil just prevents us from seeing clearly enough into the past to reach rock-solid conclusions. In sum, we have a big-brained, intelligent species whose members hunt, make more sophisticated tools than any species has before, and sometimes bury their dead in symbol-laced ways. Carrying a legacy of belongingness from their primate past, they form intense emotional connections with each other. They also eat each other. An aesthetic sense with jewelry, simple art, and pigments is occasionally evident; in some locations, Neandertals practice something akin to a bear cult, and at other places they visit deep caves in ways that hint at ritual usage of places steeped in mystery, and mysterious connection to otherworldly realms. Some advanced form of communication, most likely an early language, marks their interactions.

While it is easy, and sometimes justified, to deny to any given element in the Neandertal behavioral repertoire a religious overtone, it is much harder to dismiss the cumulative weight of the evidence. Neandertals evolved from earlier hominids, and their symbolic sensibilities evolved, too. Religious ritual in Neandertals emerged from the deeply felt and socially expressed emotions connected to these symbols.

Gaps in our understanding about the nature of these rituals are real, but the collective evidence nonetheless carries a message across the millennia: Neandertals needed each other, emotionally as well as physically, to survive. Meaning in their world was created through this emotional engagement, and at certain times and places, this process transported them beyond a strictly survival-oriented world into a new spiritual universe.

Yet the harshness of their material world may also have held them back. The expression of the Neandertal religious imagination is a limited one; the full richness of sacred ritual is yet to come in prehistory. Let's continue with the human evolutionary story, and find out what was happening in the *Homo sapiens* lineage while Neandertals walked the earth.

FIVE

More Cave Stories: Homo Sapiens

IN DANIEL QUINN'S NOVEL *ISHMAEL*, the jellyfish character insists that the world was made for jellyfish. He is so immersed in the "jellyfish worldview" that he sees no other role for the universe. Quinn's readers immediately grasp the absurdity of the jellyfish's assumption. As the story develops, they come to see that we humans, too, readily assume that the world was made for us. "Convincing people that the universe was designed with them in mind is as easy as convincing a child that candy is good for him," as one writer puts it.[1]

Our species holds a special fascination even for those who are content to see themselves as part of evolution's flow of life. Living creatures evolved on earth around 3.8 billion years ago, and primates about 70 million years ago. During the most recent 10 percent of the primate era, the hominids appeared. In this context, the longevity of *Homo sapiens* amounts to a mere geological eyeblink: about 200,000 years. As antithetical to our species' hubris as it may be, *Homo sapiens* has existed only about a sixth as long as the "failed" robust australopithecines.

Of course, *sapiens* is the only species playing the hominid-survival game now. Human history is sadly riddled with toxic crusades and bloody genocides, some carried out in the name of religion, and some quite recently. At a less violent level, human populations compete for finite resources, or force competition for resources by unequal distribution across the globe of the wealth we do have. It's obvious, and it feels entirely natural to us, that nowhere in this situation do we face other hominids as rivals. Yet such interspecies overlap was a fundamental part of our prehistory. Though I am not suggesting that outright bloodshed was part of the picture, *Homo sapiens* surely competed with Neandertals for many thousands of years. We know now that this competition lasted until about 28,000 years ago, much longer than had been thought at first.

But more surprises were in store. Excavations on an island in Asia reveal that our species shared the planet with a close cousin, quite different from Neandertals but clever in its own right, until about 13,000 years ago. This date shocked scientists. For a long time it's been known that humanity crossed a kind of cultural watershed right around 10,000 years ago, a watershed related not only to foraging and settlement patterns but also to the expression of the religious imagination. In Turkey, archaeologists have discovered a temple that dates back 11,000 years and so is one of the earliest structures that can be identified definitively as a place of worship. In that same region, the site of Catalhöyük tells us that farming began before 8,000 years ago. There, people lived in mud-and-plaster homes, grew cereals and lentils, and expressed spirituality via wall paintings and clay figures (see Chapter 6).

The temporal relationship between farming and sedentism—is hotly debated, as we will see. One thing anthropologists never doubted, though: in the millennia leading up to this time of rapid innovation, *Homo sapiens* was the lone survivor in the great pruning of the human lineage. From early diversity had come a single dominating species, honing both its survival skills and its spiritual tendencies entirely unaccompanied.

How astonishing, then, to learn that on the Indonesian island of

Flores lived a strange hominid that survived until only two thousand years before the construction of the Turkish temple. Not much taller than three feet even in adulthood, Homo floresiensis is nicknamed the Hobbit.[2] Judging by their sophisticated tools, these tiny people hunted game cooperatively, and probably coordinated their actions through language.

In 2004, news of this discovery flashed around the world, and with good reason, for its implications are stunning.[3] We are forced to recognize, once and for all, that humanity's position as *the* hominid on earth is an evolutionary novelty. And what can it mean that such small-brained hominids were sophisticated tool makers and hunters? Are we mistaken in seeing a basic linkage, however imperfectly realized, between increased brain size and increased cognitive ability in prehistory? These are only two unanswered questions among many about floresiensis.

As this book goes to press, paleoanthropologists remain locked in debate about the Hobbit: Could its small brain be due to some deformity? Were the hominids victims of microcephaly, a condition in which the brain is abnormally small because of disease? Two factors argue against this hypothesis. Seven Hobbits are now known, and all are small-brained, so we are not dealing with a single pathological specimen. Computerized analysis of one Hobbit shows, too, that the individual's brain was in some ways quite advanced and furthermore that it lacked diagnostic indicators for microcephaly. Perhaps the small size of floresiensis related to their island home; other mammals have been known to "dwarf" over time in response to specific ecological conditions associated with island living.

Less than four months after Homo floresiensis was announced, a second discovery in human evolution rocked the world of science. No new bones were involved this time. New dating techniques paved the way for a reevaluation of fossils unearthed in 1967 by Richard Leakey. Two skulls found at Omo-Kibish in Ethiopia were re-dated to 195,000 years ago, instead of the original date of 130,000 years ago. Suddenly the fossil evidence for the origins of Homo sapiens had moved closer to the DNA estimates that placed modern human origins at 200,000 years, or more. (Surprised anthropologists everywhere, including me, scrambled to update their teaching notes and in-progress manu- scripts.)

The Omo-Kibish skulls are unmistakably modern. If we gathered in an anthropology lab ten university students who knew no anatomy or prehistory, and gave them a jumble of Omo-Kibish skulls, other modern human bones, and Neandertal skulls and bones to sort, it is unlikely that they could categorize the remains by species. But none of them could fail to sort the fossils into two distinct groups. A short list for distinguishing between the two species is impressively unsubtle: anatomically modern human (AMH) skulls are no bigger than Nean- dertals', but have a high forehead, with reduced brow ridges or no brow ridges at all, and a distinct chin. As for the rest of the AMH skeleton, it is far more gracile, with bones and joints lighter and less massive than those of the remarkably strong Neandertals. Here the term "robust," imported from the robust australopithecines, takes on importance in a new context.

In less than half a year, then, our understanding of *Homo sapiens* was altered significantly by the recalibration of dates for these modern skulls from Omo-Kibish, and by the discovery of *floresiensis*. The up- shot is that the origin of our species is more ancient, and more closely connected with the presence of other hominid species, than we ever suspected.

So far, we have been talking about the evolution of modern anatomy, but even more significant for understanding origins of the re- ligious imagination is the timing and nature of *behavioral* modernity. Did modern behavior coevolve with high foreheads and gracile skele- tons? When exactly did behavioral evolution in the AMH lineage be- gin to outpace what the Neandertals were doing?

SUNGHIR AND LASCAUX

The scene is a high-altitude site on a river in Russia at Sunghir, about 120 miles northwest of Moscow. Here, 28,000 years ago, two AMH children were buried head to head. One, a girl, was about nine years old when she died; the other, a boy, was perhaps thirteen. This interment has been described as downright spectacular: "The skeletons were covered with red ocher and accompanied by extraordinarily rich and unique grave goods. Thousands of ivory beads, probably sewn onto clothes, long spears of straightened mammoth tusks, ivory daggers, hundreds of perforated arctic fox canines, pierced antler rods, bracelets, ivory animal carving, ivory pins, and disc-shaped pendants were part of the ornamentation of the burial."[4]

An adult male at Sunghir was also decorated in death, being buried with twenty-five polished mammoth-ivory bracelets on his arms, and a pendant around his neck with red paint and a single black dot. Considering this adult burial together with the double child interment, the archaeologist Randall White estimates that 13,000 prepared ivory beads were needed. By experimenting with beadwork, White concluded that the man's beads would have taken over 3,000 hours of workmanship to create, and the two children's together an astonishing 10,000 hours.[5]

The Sunghir burials reflect a social complexity, indeed a degree of belongingness, greater than those associated with even the most elaborate Neandertal rituals. Admittedly, Sunghir is not your typical AMH burial, but it shows what AMH populations were capable of under certain conditions. Most crucially, the commitment of time and creative effort involved had to have been communal in origin.[6] No single individual, indeed no single family, could have orchestrated the double-child burial or others like it. A social network of some sort, fed by emotional connection between its members, must have been firmly in place. Of course, the notion of such a network is highly predictable from everything known about prehistory, but finding it *tangibly* in the fossil record at this time is exciting!

Interestingly, the other individuals buried at Sunghir lacked deco-
rations, and apparently, any honored status. Together with the fact that
greater effort was invested in the child burial than in the adult-male
burial, this leads White to think that for the people at Sunghir, social
position was ranked and possibly inherited. His conclusion is convincing, but whether or not he has every detail correct, White's overall
theme is right on the money: social relationships are now *materially
marked*.[7]

A high-powered business executive dresses differently from a university student; a person who is gainfully employed selects different
clothing, jewelry, and hairstyle for a meeting with the boss on Friday
morning than for visiting a music club that night. In the AMH time
period, too, social identity was marked by aesthetic choices related to
personal decoration.

Until now, in my hominid-by-hominid survey, I have kept distinct
the "religion" section from all other material reviewed. Evidence relating to the religious imagination is fairly sketchy clear up until the time
of Neandertals; even there, the evidence is patchy. A division of this
sort no longer makes sense; in thinking about *Homo sapiens*, a seamless
meshing of spiritual concerns with matters of daily life is preferable
right from the start. A case study of a second AMH site, Lascaux,
helps to explain why.

Lascaux is a hominid site familiar to many readers. Adorned with
remarkable paintings, some as old as 17,000 years, this cave is cele-
brated in books, articles, and films about early AMH culture or the
origins of art. Admittedly, it is just one of a number of fascinating such
caves. At nearby Chauvet Cave is found a unique gallery of lion images;
and, at 32,000 years old, Chauvet's paintings are far more ancient than
Lascaux's. The oldest known cave paintings of all, stretching back twice
as far as Lascaux's, are the animal images found at Fumane Cave near
Verona, Italy. In Spain, for example at Altamira, are cave paintings
compelling to behold. Still, the sheer breadth and stunning quality of
the Lascaux paintings, together with the scholarly scrutiny they have

received over sixty-five years, make this site a prime choice for a case study about the spiritual side of the first art.

When writing in an earlier chapter about empathy and meaning-making in apes, I wished that readers could visit "my" gorilla family and watch the finely tuned dance of belongingness among its members. When tracing the dawning spirituality in hominids, I wondered what it would be like to see with our own eyes the face in the Makapansgat cobble, and to feel the cobble's weight in our hands. What insights would we derive from standing in the utter darkness of a chamber deep inside a Neandertal cave in France? Hands-on—or "eyes-on"—experience is worth a great deal in any of these situations. When it comes to the cave paintings at Lascaux, it is more than helpful, it is imperative to be able to visualize the subject at hand.

Ways to view the images exist but are limited. Even if we jetted to the Dordogne valley region of France, access to the original Lascaux Cave would be denied us. Discovered in 1940 by four teenagers, by the postwar years, the cave was visited by about 1,200 people a day who wished to see its treasures. Alterations to the cave entrance and floor were made to accommodate them. The up-close-and-personal encounter was exhilarating for cave visitors, but dreadful for the paintings; by 1955, they were starting to deteriorate visibly as a result of high levels of exhaled carbon dioxide in the air. In 1963, the cave was closed to the public, though visiting scientists are sometimes allowed brief entry by prior arrangement.

In recognition of its national treasure, the French government designed Lascaux II, a scrupulous mock-up of major parts of the original cave. Touring this replica is the closest that most of us can come to the original experience. A second-best substitute is a good video documentary or, conveniently available on the Internet, a virtual tour. I strongly urge you to take some time to look at the Lascaux paintings now, and to continue viewing them as you read.[8]

Why is the visual experience critical? Anyone could be forgiven for imagining that *Homo sapiens's* early paintings would be crude, like a child's rendering, a decent "first try." After all, when *Homo habilis* first struck stone against stone to make cobble and flake tools, the result was far from a model of technological efficiency or aesthetics. The cognitive breakthrough was huge (no other creature, to our knowledge, had done this before, and assuredly it took skill), yet the products themselves were on the crude side. Wouldn't it make sense that the first cave art of many thousands of years ago might be simple precursors to what came later, when artistic skill and experience had blossomed?

Yet the Lascaux paintings are no childlike, tentative attempts to draw the world. Parading across the cave walls are cavalcades of large animals, horses, and deer as well as bulls, each rendered in glorious and colorful detail. Motion is painted right into these images; observers gain a sense of the animals as active beings in the world. The ancient artists even understood the concept of perspective. When painting an animal high atop a wall, they distorted the proportions so that, when seen from below, the creature would look realistic.

From the two-dimensional representations available to us, we can only try to imagine what it feels like to stand inside the real Lascaux and gaze at the images. David Lewis-Williams, an expert on prehis-

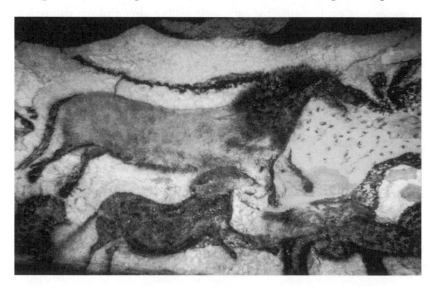

Horse images painted 17,000 years ago at Lascaux Cave.   *Art Resource, NY*

toric art, writes, "a prominent American archaeologist, who was granted 20 minutes in the cave, told me that the first half of his allotted time was rather wasted because, overcome by the wonder of it all, he viewed the art through a curtain of tears." Because Lewis-Williams himself was granted "generous access" to Lascaux, he is the ideal person to donate his "eyes on" experience to us.[9]

Arguably, the most powerful of the cave's seven sections is the Hall of the Bulls. The first-discovered and most accessible of the chambers, the Hall of the Bulls is packed with large colorful paintings. In addition to the bulls, deer, and horses is found a fantastic-looking animal, sometimes called the Unicorn by art prehistorians. As Lewis-Williams points out, the nickname seems inapt, because the animal is depicted with two horns. That it is a fantastic creature is beyond doubt. Collectively, the images crowding the Hall of the Bulls show us that the Lascaux people could, time and time again, render their subjects with a mix of the realistic and the imaginative.

Using brushes and other tools made from organic substances such as wood, skin, and horsehair, Lascaux people painted with a palette of pigments ranging from bright red and yellow to black. Frequently the artists, when choosing a location to paint an animal image, took into account the cave's natural contours: hollows and curves became part of the animals' anatomy just where appropriate. Color and naturalistic detail combined to make many of the images look strikingly accurate, even to us today.

Sometimes, though, reality is left behind. Stags are given curiously elaborated antlers. The so-called Unicorn seems to be a jumbled-together animal: perhaps a rhino, a bison, a big cat, and a horse? Here is no prehistoric creature, but some mysterious animal of the mind.

Perhaps the first thing that strikes a visitor entering the hall, judging from what Lewis-Williams reports, is the art's very magnitude. At about seven feet in length, for instance, the Unicorn is scaled to the large size of the chamber itself, which measures nearly thirty feet across in the painted section. In the same way that the huge number of decorative beads at the Sunghir burials clued us in to a social network there, the size of the paleoart in this hall hints at the social nature of life at Lascaux: "The sheer size of the images suggests that they were communally made. People must surely have co-operated in the preparation of paint, construction of scaffolds, outlining the huge images, and then the application of the paint, even if one, or a few, highly skilled people directed the work."[10]

But art and sociality are linked even more deeply than this; *all art is social because all art is symbolic.* Lewis-Williams understands this: "It is not possible for someone to depict an animal without in some way giving resonance to some part of its symbolic associations, and those associations are socially created and maintained."[11] A painting made by a man or woman working alone, whether in a dark Ice Age cave passage or a light-flooded present-day loft in Manhattan, is just as much a social creation as any piece of art that was made by multiple artists who daubed paintbrushes on the same canvas.

And from its inherently social nature, other roles for art may be born. The Hall of Bulls is so big that entire social groups could have congregated there after the images were first created, to view the art or perform ritual acts. It is almost impossible to envision life at Lascaux, in fact, or life at any of the other art-rich caves in Europe during this period, in the absence of ritual.

Clues to ancient ritual at Lascaux are found in the room next to the Hall of the Bulls, the Axial. Peppered with animal images on the walls and the ceiling, the Axial ends in a tight area called the Meander, where passages are quite narrow. The Meander's centerpiece is the Falling Horse, described by Lewis-Williams as "a miracle" of AMH artistry. Legs up in the air, the horse is depicted upside down, but because of the shape of the passage, its entire form is not visible from a single vantage point. For a view of the entire horse, cave visitors must crouch down and then shift position a bit. In a small niche in the wall across from the Falling Horse rest three flint blades, painted red. And at the bottom of a shaft in another area of Lascaux is a much-analyzed image of a strangely primitive, bird-headed human figure. Depicted at an odd angle and next to a bird standing on some type of a pole or stick, this figure has an erect penis.

How can we understand the upside-down horse, the flint blades, and the bird-human hybrid next to a pole? It helps to know that parts of the cave were visited by Lascaux people in darkness or near-darkness. Ancient lamps found in the cave confirm that some human-made light sources were available, but these would have only weakly illuminated the blackness. In these dark places more than anywhere else—just as with Neandertals in their caves, but here with art all around—a burgeoning sense of the spiritual was likely to emerge in *Homo sapiens*. And there is another difference between what *Homo sapiens* did in the art caves and what Neandertals did in caves like Regourdou: in places like Lascaux, shamanism was born. It is by exploring shamanism that the seamless relationship of art with an emerging religious imagination can be best understood.

SHAMANISM[12]

Shamanism is so widespread across traditional cultures in the modern world, and so key to the human adaptation throughout the ages, that anthropologists have studied it from every conceivable angle—and not only in the manner of observers, or even as participant-observers in other cultures. In California, the anthropologist Michael Harner has established the Foundation for Shamanic Studies with the aim to study, teach, and preserve shamanism. Along the way, the foundation trains upward of 5,000 people a year around the world in shamanic techniques.

The word "techniques" points us in the right direction, because virtually all scholars of shamanism emphasize that it is not a religion. It is a way of approaching the relationship between this world and a spirit world. In other words, it is a method. Specifically, the method of shamanism is deeply bound up with altered states of consciousness. The core experience for a shaman involves traveling to the realm of spirits while in some kind of trance state. Once in the spirit world, the shaman endeavors to carry out a helpful act, perhaps healing a sick person or finding out information crucial to the group's survival, such as the location of prey animals to be hunted.

Harner, who has lived among traditional people in many corners of the globe, practices shamanic healing himself. Acting something like a cross-cultural interpreter, he helps those of us with no direct experience of shamanism to understand the shaman's reality: "Shamans don't believe in spirits. Shamans talk with them, interact with them. They no more 'believe' there are spirits than they 'believe' they have a house to live in, or have a family. This is a very important issue because shamanism is not a system of faith. . . . Shamans talk with plants and animals, with all of nature. This is not just a metaphor. They do it in an altered state of consciousness."[13]

Though shamanism is not religion, it shares something with religion as I defined it in Chapter 1. Shamanism is about *action* (in this case, through altered consciousness), or more specifically, about *transforming* the world through action. This is something that Karen Armstrong would understand, given her acute grasp of the relationship of religion and compassionate action recounted in *The Spiral Staircase* (see Chapter 1).

The notion of transformation shines out through the words of a person who spent his life as a shaman. In the classic book *Black Elk Speaks*, a Lakota Indian man named Nicholas Black Elk tells the story of his life as a shamanic healer.[14] Experiencing visions from the spirit world, Black Elk acted in ways that powerfully changed his own life as well as the lives of others around him. His first vision, a glorious one, came when he was nine. Voices summoned him, and when a cloud came down from the sky, it transported Black Elk to another world. There he encountered all sorts of creatures, including twelve black horses with manes of lightning, and thunder in their nostrils. He met the Powers of the World (six Grandfathers of the World), who gave power to him. Much more happened during this vision than can be told here.

When he returned to himself, Black Elk was sitting in his family tepee. His parents' experience of the twelve intervening days involved caring for their very ill son, who was "lying like dead all the while." Reflecting on the reality he experienced during those dozen days, Black Elk writes, "I could see it all again and feel the meaning with a part of me like a strange power glowing in my body; but when the part of me that talks would try to make words for the meaning, it would be like a fog and get away from me."[15]

I love this passage, which expresses so eloquently what it is like to *feel* meaning: to experience, in an embodied way, a connection with something bigger than oneself, a connection whose precise contours cannot be articulated to others (or even fully to oneself). We are in Martin Buber's realm, where a human being's relationship with God is wrapped in a cloud but reveals itself, and lacks but creates language.

Black Elk's vision changed him. He began his new life as a shaman, and remembers his first experience of healing another person: "Everything was ready now, so I made low thunder on the drum, keeping time as I sent forth a voice. Four times I cried 'Hey-a-a-hey,' drumming as I cried to the Spirit of the World, and while I was doing this I could feel the power coming through me from my feet up, and I know that I could help the sick little boy."[16]

Such first-person tales, and anthropological accounts, too, make compelling reading and identify important themes and patterns in shamanism. But what makes prehistorians so sure that shamanism was practiced during early AMH days in Europe?

Ethnographic analogy with the deeply emotional, experiential, and shamanic nature of religious expression in hunter-gatherer cultures around the world can be a useful starting point. We are projecting behavior back in time to ourselves, after all—to an early incarnation of our own species, already equipped with big brains, language, and an elaborated version of primate belongingness. From the time of our apelike ancestors we can see the legacy carried forward right into the Lascaux cave. Who, after all, is more adept than a shaman at expressing the fullest possible empathy? Who is more skilled at meaning-making than someone who communicates between the human world and its

The bird-human hybrid figure and its shaft hint at shamanism at Lascaux Cave.
Art Resource, NY

supernatural counterpart? What method (for, as we have seen, shamanism is a method and not a religion) works without a set of rules to follow? And who can experience visionary transformation without a deep imagination? Ancient primate capacities are now transformed, in *Homo sapiens*, into something unrecognizable to any earlier primate.

Yet at Lascaux, more specific and localized support for claims of shamanism can be uncovered.

Consider again what Lascaux is all about: imaginatively drawn animals, many depicted floating in space rather than anchored in an ecologically realistic context, and some that were painted in areas dark and relatively inaccessible, but nonetheless used over and over again by the artists. Some of these images may be shamanic in origin. It is very tempting to interpret the shaft drawing at Lascaux in shamanic terms: a crudely rendered male human with a birdlike head and a prominent

Ancient Indian petroglyphs: the procession panel at Comb Ridge Canyon, Utah. The guide, carrying a shaft, is reminiscent of a figure at Lascaux.   *Alan Fogel*

erection leans back, next to a bird perched on some kind of pole or stick. The figure's crude quality sets it apart from the sophisticated animal images. Does it represent a shaman in an ecstatic trance? Birds and pole-like staffs are both often associated with shamans, so the suggestion is perfectly plausible.[17]

According to a German researcher, an Ice Age star map is located right near the man and bird, which are themselves positioned near an image of a bull. Michael Rappenglueck says that the eyes of the bull, the man, and the bird represent three stars (Vega, Deneb, and Altair) known collectively as the Summer Triangle. Even for modern city-dwellers, these stars are easily visible to the naked eye. For the Lascaux people, not saddled with smog or urban light pollution, they must have been vivid indeed during summer months. If Rappenglueck is right about this star map; about his claim of a second map at Lascaux, this one of the star cluster we know as the Pleiades; and in his conclusion that the sky was "full of animals and spirit guides" at this time, the cave's celestial representations offer us further clues to the spiritual nature of our ancestors.[18]

Taken together with ethnographic analogy, then, features of Lascaux Cave support the view of Brian Hayden and David Lewis-Williams that shamanism was part of life in the European art caves of this period. (Indeed, Lewis-Williams argues for trance-induced hallucination at Lascaux.) Study of the Lascaux paintings lets us peer back through the millennia at the tangible marks of a human society steeped not only in symbols, and in symbol-based ritual, but also in spiritual ritual.

In sum, then, by the time of Lascaux, and probably of Sunghir as well, human beings were deeply emotionally engaged not only with one another but also with some aspect of the spiritual world. Complex, symbol-mediated behavior was emerging at a new pace and on a bigger scale than ever before; life was changing everywhere in Europe, and some of those changes related directly to the origins of religion.

THE AMH WORLD IN BROADER PERSPECTIVE

For hominids before the Neandertals, tools and how they were used provide the most direct way to assess behavior and culture. With *habilis* those tools were cobbles and flakes; with *erectus*, hand axes. Belongingness and emotion are present in the lives of these hominids, but cannot readily be assessed by examining tools or by mapping the long-distance travel routes of the *erectus* groups that, generation by generation, trekked slowly out of Africa.

With Neandertals comes a turning point of sorts. Neandertal burials, even with their symbolic components such as grave markers, cannot all on their own point definitively toward spirituality. But taken together with everything else we know about Neandertals, ranging from the Regourdou bear rituals to the flowering of an aesthetic sensibility and their use of deep caves, the burials help build a convincing case for thinking that belongingness expanded into a spiritual sense in these hominids. That spirituality may have been expressed in a "patchy" way, in only some places at some times, or far more robustly; the evidence does not speak to that. When *Homo sapiens* colonized western Europe, their cave sites are permeated with spirituality. But here, too, we need a collective picture to interpret. What else do we know about *sapiens* of this time period, apart from the Sunghir burials and the Lascaux art?

At the time of Sunghir and Lascaux, *sapiens* people made a wide range of bone, antler, and ivory tools, which enabled them to accomplish more tasks, more efficiently, than ever before. Fishhook technology, for instance, meant higher fishing yields—though, as we will see, earlier *sapiens* people feasted from the sea as well. Bone needles made the fashioning of garments that much easier, though clothing had probably been in use for some time. (According to an unconventional and perhaps mildly repulsive source of information—the evolution of body lice—clothing may be as old as 70,000 years.[19])

Beauty mattered to these tool makers. Laurel-leaf points were fashioned as a regional specialty for a few thousand years. Finely crafted, indeed exquisitely delicate, these points are so long and thin that they may have been entirely useless for any practical task at all. Scientists joke that they may have been the first luxury items.

But survival mattered. People now constructed hutlike shelters, some complete with hearths; natural caves and rock shelters were no longer the only hominid homes. Of all the game animals roaming their world, the mammoth provided the most resources to these hunter-gatherers, and not only in terms of calories from protein and fat: the bones, the tusks, and probably the hides became critical resources for prehistoric construction crews. At Dolni Vestonice in the Czech Republic, a site important in the prehistory of art (see below), people used mammoth bones and, in one case, clay, to construct huts. Mammoth bones also framed a sturdy house at a place called Mezhirich on the steppes of Ukraine, though this was later, at about 15,000 years ago. Ten men working together would have taken six days to manufacture the Mezhirich dwelling, archaeologists estimate.[20]

Groups of people cooperated in new ways, and at the same time, this ancient, within-group belongingness began to expand outward. Materials like amber and shell begin to turn up in places far from their origins, an indicator that regional trade networks were growing in importance. Different groups of *sapiens* probably communicated with each other through complex and systematic patterns of exchange. Perhaps this is partly why group identity, through personal, style-specific decoration and perhaps other customs, now becomes vitally important. Using dress and jewelry to send the message "I belong to *my* group and not *your* group" is a robust human tendency, and an ancient one.

What we see then, is that ancient *Homo sapiens*, the tool makers and hut dwellers whose artifacts are so well known to science, can be fleshed out as living people. At the end of the day it is the commitment to *living beyond survival* that is most arresting and noteworthy in *sapiens's* behavior. It's no exaggeration to say that some of the art and ornamentation from this period is breathtaking. Completely different in scale from the Lascaux paintings, three figurines from the Hohle Fels Cave in Germany attest to the magnificence that can be captured even in works just an inch or two tall. The miniature ivory carvings of a horse, a diving waterbird, and a half-man-half-lion, made about 33,000 years ago, include beautifully rendered details.

Hayden, Lewis-Williams, and other scholars of the evolution of shamanism surely must welcome the Hohle Fels finds. In an interview with *National Geographic*, the figures' discoverer, the German anthropologist Nicholas Conard, explicitly links two of them with shamanism. The blended man-lion figure fits well with shamanism, which for hunter-gatherers often involves human-to-animal transformations; as for the bird, waterfowl are frequently helper guides to shamans in reaching the spirit world.

Clay figurines appear to be common at this time; AMH populations in the Czech Republic even fired the clay. From Dolni Vestonice comes a 26,000-year-old "Venus" figure, the oldest known baked clay figure in the world. A mini-statue under five inches tall, it is all breasts and hips. I don't mean this description as a crude idiom: hundreds of Venus figurines are known across Europe, and as I will soon describe, they have supremely exaggerated female characteristics.

The home of Ice Age hunter-gatherers who warmed themselves around hearths in their huts, Dolni Vestonice is now rich in the remains of art objects. Carved animal and human images, and items of personal ornamentation such as necklaces, litter the site. Two kilns and thousands of fired fragments attest to the importance of making clay figurines there. Apparently, the fired objects were made at temperatures around 1,300 degrees Fahrenheit. Because the figures show numerous thermal cracks, one scientific analysis essentially tsk-tsks at the Dolni people's inability to master this method. But analysis of the fragments, and modern-day experiments in firing techniques, suggest that it's far more likely these artists put their unfired clay into the kiln in such a way as to *ensure* breakage, or even explosion.[21] Such intentional breakage virtually screams "ritual" across the millennia.

Dolni Vestonice's huts, elaborate burials, and extensive art objects make this location a five-star site for prehistorians. At the very center of attention sits the Venus figure. Variation is the watchword when thinking about Venus figurines. Some are made of stone, some of ivory, some of bone, some, as we have already seen, of clay. Some were naked, others clothed. Artists did consistently highlight female sexual anatomy—breasts, buttocks, and vulvae as well as enlarged stomachs—though in patterned ways that differed across regions.

In graduate school, I was taught that these statues were all about fertility, perhaps even the material evidence of a fertility cult. In her book *The Myth of Matriarchal Prehistory*, Cynthia Eller convincingly debunks this idea. Not one of the Venus figures is pregnant or lactating, nor do any cradle a babe-in-arms.[22] If this is symbolism about fertility, it is a strangely veiled sort. (It is possible, but highly improbable, that prehistoric people understood the link between the "reproductive fat" on a woman's body and her ability to conceive successfully.)

Research by the anthropologist Olga Soffer shows that there's more to these figurines than breasts, buttocks, and hips. Though burials such as the ones at Sunghir involved clothed and decorated bodies of both males and females, there's no such gender equity in the statues of this time period. Only female figurines are ever clad, sometimes in a kind of string skirt. Some wear carved hats or caps, upper-body bandeaux and belts. Soffer and her colleagues make an interesting argument based on this gender disparity. Integrating information from a number of sources—the ancient art itself as well as analogy with living peoples—they conclude that women were the weavers and producers, in real life, of the garments depicted on the Venus images. This clothing production probably brought to women real prestige: "The exquisite and labor-intensive detail employed in the depiction of the woven garments worn by one group of Venuses clearly shows that weaving and basket-making skills and their products were valued enough to be transformed into transcendent cultural facts carved into stone, ivory, and bone."[23]

Here a long-ignored role for women grabs attention front and center. What does this mean for ritual and religion? One fascinating thing is that these prehistoric artists rarely carved an unambiguously male form. Why the almost exclusive focus on the female form? It is tempting to think about some kind of ritual-based cult, though not necessarily one bound up with fertility. Any specific scenario must be viewed with caution, but keeping in mind the general picture of ritual-based *sapiens* life we are drawing here, if we think about a link between the art's meaning and ritual we are probably headed in the right direction. Both Eller and Soffer support the linkage of these figurines with ritual, possibly religious ritual.

The regional clustering of Ice Age art is quite striking and a clue, perhaps, to an ecological "push factor" for this creative expression. Hayden links the complex art and ritual we are discussing with the availability of game. In southern France, northern Italy and Spain, and the Ukraine, there existed a common ecology: "Other areas apparently either had lower animal densities or animals that did not migrate to the same extent, or their migrations were not as confined, and were more dispersed, making it more difficult to capture large quantities of them."[24] Perhaps this abundance of calories on the hoof simply freed people to follow their creative urges. That Regourdou, the Neandertal burial-and-ritual site described in the preceding chapter, is in this same area (less than a mile from Lascaux) makes Hayden's suggestion all the more intriguing.

How do these behavioral shifts in *sapiens* living in Europe—in spheres ranging from production of large-scale communal art and jewelry to shamanism—relate to the behaviors of *sapiens* living earlier in time or outside of Europe, and indeed, those of Neandertals themselves?

BEHAVIORAL REVOLUTION?

It's a boilerplate conclusion in anthropology: Lascaux's splendid art, and Sunghir's elaborate burials, represent the finest examples of a behavioral revolution enacted by *Homo sapiens* living in western Europe after about 40,000 years ago. By adopting the term "revolution," advocates of this view wish to point to something beyond a mere quickening of the pace of cultural evolution, indeed to "an explosion in explicitly symbolic behavior and expression" as the archaeologist Paul Mellars puts it. The argument goes like this: Certainly, the occasional scrap of art or other evidence of symbolic creativity can be identified before 40,000 years ago, but only at this date, and only in western Europe, was a genuine boundary crossed. This line of thinking doesn't necessarily hinge on the revolution *beginning* in Europe; it is a nuanced argument, one long accepted and best summarized elsewhere.[25]

But a theoretical sea change may be under way. Smashing up against these accepted ideas is evidence from a place called Blombos in South Africa. Perched cliffside above the Indian Ocean, this cave was home to AMH people whose actions defy any notion of a late behavioral revolution centered in western Europe. At around 70,000 years ago, Blombos people engraved onto two pieces of ocher a geometric pattern of quite some intricacy. The archaeologist Christopher Henshilwood believes that the pattern of cross-hatchings and lines on the ocher amounts to far more than the doodling you or I might do to pass the time waiting on hold during a telephone call: "We don't know what they mean," he says, "but they are symbols that I think could have been interpreted by those people as having meaning that would have been understood by others."[26]

At Blombos is a foreshadowing of what was to come in the Sunghir double-child burial, for these seaside residents ornamented themselves with shell beads. At about 75,000 years ago, forty-one tiny mollusk shells, grouped into clusters, were pierced in a uniform way. This deliberate piercing, together with certain wear patterns on the shells (wear from the material used to string them and from the action of the shells rubbing against each other), tells us that some type of jewelry was made at Blombos. Pigment was used, too; if you guessed red ocher, you are right.

Personal identity mattered at this site, where the inhabitants made a good living day by day. In reconstructing ancient life at Blombos, Henshilwood conjures up people living off the "bountiful larder" of the sea as well as the land. People speared fish and collected a range of edible species, from periwinkles to seals and dolphins. Perhaps the ease of

foraging, with time freed at Blombos for leisure pursuits—the making
of images and of jewelry—is an early instance of a situation that recurs
later on when cave paintings were created in Europe's corridor of high-
density game migration.

East Africa, too, yields evidence of symbolism early on, though not
as early as Blombos. At a rock shelter site in Kenya dated to 40,000
years ago and called Twilight Cave (in the local language, Enkapune Ya
Muto), people pierced fragments of ostrich eggshells, turning them
into beads. The beads were tiny, only a quarter-inch in diameter, and
must have taken great care to fashion. If ethnographic analogy is a valid
guide, the beads may have been used to cement social relationships.[27]
Once again, jewelry and belongingness are linked.

The African archaeological record may well have numerous riches
to yield in this story of early symbolism, not only between the time of
Blombos and that of Twilight Cave, but before and after as well. Cave
painters were at work on this continent. Conflicting dates exist for the
art at Namibia's Apollo 11 Cave, but the country's National Museum
puts them at about 26,000 years ago, if not earlier. Animal images on
slabs in the cave resemble today's rhinoceros and zebra. Apollo 11 Cave
is nowhere near as well-known as Lascaux, but the beauty represented
here and at other African sites deserves an appreciative look too.[28]

Intriguingly, the current contender for oldest symbolic behavior by
AMHs comes from neither Europe nor Africa, but from Israel. At
Qafzeh, *sapiens* skeletons dated to 90,000 or even 100,000 years ago are
stained red—yes, ocher again. Archaeologists believe that the Qafzeh
people may have used this red pigment to represent death.[29] A conser-
vative approach requires a caveat here, because not a single clue has survived as to what a concept of death might really have been to people
living so long ago. Just as we cannot truly grasp what, if anything, death
means to living apes when they act distressed at the body of a dead
companion, we cannot know what red ocher means at Qafzeh. But
Homo sapiens are not apes, and it's a good bet that at Qafzeh ritual practices of some sort surrounded death.

The representational use of red ocher in any systematic way is significant, then, especially in combination at Qafzeh with pierced shells used as ornaments, and the placing of deer antler at one grave to mark a burial. The emerging picture of life at Qafzeh gives us, at the least, a baseline for human behavior at 100,000 years ago. In this time period, *sapiens*, like their cousins the Neandertals, are neither trapped entirely in the present nor entirely consumed with issues of day-to-day survival.

Armed with the information from Qafzeh and from the African sites, the "European revolution" scenario can be challenged in favor of a deeper and wider behavioral continuity. Evidence from Down Under plays a role, too, because bones at the site of Mungo show that *sapiens* had reached there by about 50,000 years ago—and how else to colonize Australia except by boat? If Asian *sapiens* cooperated in groups to make watercraft at around this time, for journeys to Australia, significant revisions to a Eurocentric behavioral revolution are in order.

From Australia come the famed fossils "Mungo Woman" and "Mungo Man," two members of a lakeside community of hunter-gatherer-fisher people. Mungo Woman's body was burned in what may have been the world's first intentional cremation (after which her remains were buried). The man's bones were graced with red ocher. By 40,000 years ago and probably earlier, then, symbolic treatment of the dead had reached this distant continent.[30]

AMHS AND NEANDERTALS

Leading voices in the behavioral-continuity camp, those of the anthropologists Alison Brooks and Sally McBrearty, confidently proclaim that symbolic behavior stretches back at least 200,000 years.[31] Given my own work on the evolutionary continuity of language and culture, indeed given the theme of this book, it will come as no surprise that I stand firmly within this camp. I want now to bring Neandertals back into the equation, because whatever the timing of behavioral modernity turns out to be, the story of *sapiens's* evolution is completely bound up with the story of the Neandertals.

The long-perceived cognitive and behavioral gap between AMH life and Neandertal life is narrowing with every new discovery. In places like Arcy-sur-Cure in France, Neandertals made refined blade tools to rival any produced by early AMHs. They successfully hunted fearsomely large game. Yet, as we have seen time and again, it isn't this type of behavior that speaks to us most from our prehistory. What invites us to think of Neandertals as creatures who shared our sense of beauty, of mysterious connection with forces in the universe, comes not from technology or hunting but from the ways they expressed belongingness and spirituality.

Even so, early AMH populations began to pull away, achieving a new level of behavioral complexity by around 50,000 or 40,000 years ago. By 28,000 years ago, Neandertals were no more. What was the nature of the interaction between these two species? Did AMH behavior contribute directly to the extinction of the Neandertals?

Early *sapiens* and Neandertals overlapped during at least two distinct periods of time. The anthropologist Ofer Bar-Yosef refers to a "Middle East bus station" in the general area around Qafzeh, Israel, and its neighboring *sapiens* site of Skuhl. AMH populations inhabit the region first, but by about 60,000 years ago, Neandertals live right next door (a hundred meters away) in Kebara Cave. Judging from the positioning of a famous Neandertal skeleton nicknamed Moshe, Kebara dwellers practiced intentional burial, though not in as elaborated a way as at Regourdou in France.

Neandertal and *sapiens* groups crisscrossed the same region of the Middle East, but it isn't entirely clear whether the two species came into direct contact. They lived somewhat parallel lives but had no face-to-face confrontations, as far as we can tell. In western Europe, though, the story is—or becomes—quite different.

The burials at Regourdou and La Ferrassie, and at numerous other sites as well, tell us that Neandertals flourished in Europe for many thousands of years. More than that, they had the whole area to themselves, at least with respect to other hominids. Only around 40,000 years ago did AMHs begin to colonize these regions of Europe. Global migration patterns of the two species are complex, determined in large part by climate changes related to the Ice Age. As it turned out, France, in particular, was a kind of prime overlap zone between the two species. For instance, at Vienne, AMHs and Neandertals alternated short-term usage of a single cave 40,000 years ago.[32]

Whereas in the Middle East the species were separate, in France Neandertals must have met *sapiens* up close and personal. The question everyone wants answered is, Just how personal *was* the contact? Did *sapiens* ever interbreed with Neandertals? Strictly speaking, no two species are supposed to be able to interbreed, at least not with the result of healthy, fertile offspring. When a horse and a donkey mate, a sterile animal results—a mule. And judging from recent analyses showing that Neandertal DNA is distinct from our own, it is accurate to speak of Neandertals and *sapiens* as separate species.

Even within the dual-species framework, the notion of a Neandertal-*sapiens* mating pattern has not faded from popular imagination. New techniques can now address this issue with some finesse. When DNA was successfully extracted a few years ago from a 29,000-year-old skeleton of a Russian Neandertal infant, scientists proclaimed it distinct from our own—too distinct for Neandertals to be ancestral to *sapiens*. This discovery is just one in a series to yield identical answers: *sapiens* evolved from an earlier form of *Homo*, not from Neandertals. Several candidates for this ancestor, some variant of a post-*erectus* hominid, are available. (Here's where a good textbook in biological anthropology comes in handy.)

But has a strange finding from Portugal muddied these waters? At a site called Lagar Velho, anthropologists unearthed the skeleton of a male child thought to have died at about age four. According to the by-now-familiar markers of red ocher and pierced shells, this prehistoric burial was a ritual one. In themselves, these indicators are unsurprising, given that the boy died at about 24,500 years ago. But the skeleton itself would flummox every one of our hypothetical bone-sorting college students. Though the Lagar Velho child has the teeth and the chin of *sapiens*, his bones are remarkably robust, indicative of a powerful

build. Indicative, that is, of a feature squarely associated with Neandertals—in a child living thousands of years after the Neandertals sup-
posedly became extinct as our nonancestors.

The fun of studying prehistory comes with puzzles like this one,
where the pieces refuse to fit together neatly. The anthropologist Erik
Trinkaus is certain the Lagar Velho boy is a hybrid, the product of
mating between a Neandertal individual and a *sapiens* individual. For
Trinkaus, the Vienne cave, the one shared by Neandertals and AMHs
in quick alternation, only supports the idea that the two species social-
ized and interbred. Yet . . . there's that genetic evidence from DNA
analysis.

There's no arguing that the bulk of the evidence points to a nonre-
productive mixing of the two species. Possibly this mixing included
some sharing of cultural skills. As we saw in the last chapter, production of the most refined Neandertal tool kits came about either
through aping of what AMHs were doing or through a blossoming of
skills within Neandertal groups themselves.

An "arms race" of cultural achievement may have been stimulated
whenever Neandertals and *sapiens* had to compete for resources—
food, shelter, raw materials for tools—in Europe. An early version of
keeping up with the Joneses may have spurred cultural change in an
upward spiral. This idea makes sense to me because it is rooted in powerful *social* forces (discussed further in Chapter 7), not only in climate-
based pressure for change, or in other factors of ecology alone.

Imagine how overwhelming it would be for twenty-first-century
sapiens to colonize a new environment (on earth or in space) only to
discover that higher-tech creatures more clever than we, were already
thriving there. A profound identity crisis for humanity would be the
result. But, assuming a degree of peaceable interaction and cross-
species communication, after some time, our own weaker technology
would almost certainly ratchet up: we'd copy alien ways of doing
things, or internal social pressures might result in innovation within
our own groups, or a mix of the two might occur.

Still, an intriguing idea is not the same as an accurate one; as is so

often the case with behavioral prehistory, more questions arise than answers here. Whatever happened to the Neandertals, it seems to have
happened only quite slowly rather than during some cataclysmic event.
No bloody war with *sapiens* caused the Neandertals' extinction; outright competition for resources between the two species was prolonged. Even a tiny behavioral advantage on the part of *sapiens* would
have accumulated over the generations to give them (us!) an edge.
Quite possibly language was involved, as complex linguistic skills may
have kicked in around this time, unique to *sapiens*.

To an unprecedented degree, Neandertals were capable of empathy, co-constructed meaning, rule-following, and imagination. Quite
possibly that imagination soared to the heavens, or to their Ice Age
equivalent; Neandertals seem to us, looking back on them, poised on
the verge of centering their lives around the expression of the religious
imagination. They were not, however, *as* empathetic, *as* good at
meaning-making and rule-following, *as* imaginative (or indeed as spiritually imaginative), as early *sapiens*. And for this they paid the ultimate
evolutionary price.

We have just taken a short tour through the anatomical, behavioral, symbolic, and religious origins of our own species. The fossil and
genetic evidence strongly suggest a single origin point in Africa, with
evolution from earlier African hominids; dispersal from Africa in a
"second wave," of *sapiens* (following *erectus* as the first wave); and prolonged overlap between *sapiens* and other hominids right until the tail
end of the exclusively hunting-and-gathering period. After millions of
years for which only tantalizing hints exist, the archaeological record
of *Homo sapiens*, both in Europe and elsewhere, points to expression
of religious ritual through the social forces of art, group identity,
and shamanism.

"What" and "when" questions have dominated this chapter and the
previous two. We now return to our time line, to explore the religious
imagination expressed during the period when *Homo sapiens* first invented farming and settled in cities, and then reenter the realm of explanation as well as description. Why are we the spiritual ape?

SIX

Transformations in Time

NEANDERTAL HUNTERS fan out across the frozen steppes of Europe, chasing down a mammoth. Spears at the ready, they close in and prepare to make the kill.

If any single image of human evolution is centrally lodged in the popular understanding, this one is it. It has everything: drama and danger; the quest to tame nature (in the guise of a fierce beast) by means of culture (a hominid-made weapon); even males banding together to ensure survival of the women and children.

In reality, the pairing of male hominids with heroic events is a shaky foundation upon which to reconstruct our past. The anthropologist Olga Soffer and her colleagues note that a focus on males leaves us "in the proverbial outer darkness as to what the Paleolithic 'silent majority'—the mates, children, and parents of such brave prehistoric men—may have been doing with their lives in addition to admiring and assisting them."[1] Soffer's own research shows that Ice Age women's skilled weaving and basketmaking was prized enough to be incorporated into the art of the day.

Gender balancing of this sort is much needed, but deeper layers of our prehistory, too, cry out to be revealed. More than anything, as we have seen, the story of human evolution, especially the evolution of the human religious imagination, is about *transformation through ever-more-complex belongingness*. And it is to explaining that transformation over time that I devote this chapter.

First, I survey examples of the expression of spirituality in two time periods beyond that of Ice Age hunter-gatherers: when humans first began to live in settlements, and then, leaping ahead in time, when new principles of religious expression began to cohere around the world during the Axial Age. Because the prehistoric era I know best ends at the time of the Lascaux people, I discuss these two later periods only briefly. My hope is to give just enough information to convey a flavor of what came next in the chronology of the human religious imagination. From there, I move into the realm of theory and explore exciting new ways to understand human emotional relating and its evolution and how they may help us to understand changes over time in religious behavior.

ANATOLIAN WINDS OF CHANGE

The place: atop a hill in what is now southeastern Turkey, near Syria, a site now called Göbekli Tepe rises up in the midst of a great plateau. The people: hunter-gatherers rather than farmers, the Göbekli Tepe people lived off the land, just as humans and their ancestors had all through the millennia. The time: 11,000 years ago, a mere tick of the evolutionary clock beyond the time of the Hobbit, that astonishing mini-hominid (*Homo floresiensis*) who co-existed with modern humans in Indonesia.

If only we could time-travel back to view the actions of those people, we would encounter an expression of the human religious imagination different from anything before it, for at Göbekli Tepe, people worshiped at a monumental temple. Two things are central to understanding Göbekli Tepe: the scale of its structures and the variety of its art.

Fifty-ton blocks of stone were moved around at Göbekli Tepe, a dead giveaway that its inhabitants worked cooperatively. People hauled huge stones up the hill with no domesticated animals to help with the back-breaking labor. Buildings set into the ground, almost under the ground, were supported by enormous T-shaped columns adorned with animal figures. But more than animals are depicted here. Archaeologists at the site have found male figures with erect penises; reliefs with sexualized scenes; and, incised on the floor of one building, a woman who some say is depicted in the process of giving birth.

In ancient times, people flocked to Göbekli Tepe from the surrounding region to participate in religious ritual. As we will discover, this is the conclusion embraced by archaeologists, and it seems as certain a conclusion as could be reached in the absence of written records. The temple worship at Göbekli Tepe is situated in time precisely halfway between the religious-oriented behavior at Lascaux and that in ancient Egypt: 6,000 years after Ice Age people in France adorned cave walls with glorious animals, images that hint so tantalizingly at a relationship with the world of the sacred, and 6,000 years before the inhabitants of the Nile region erected their great pyramids, symbols of belief in an afterlife from a culture saturated with the sacred. That the Göbekli Tepe people were heavily invested in religious expression so long ago is a significant finding to add to our time line of the human religious imagination.

A millennium and a half after Göbekli Tepe, fascinating things began to happen at a Turkish site called Catalhöyük. About 8,000 people settled there in densely clustered mud-brick houses. The Catalhöyük people grew wheat, barley, lentils, and peas, and they herded sheep and goats. No longer living as hunter-gatherers, as did the Göbekli Tepe people and every hominid population before them, the Catalhöyük people domesticated animals and cultivated crops. And it seems that they structured their lives around the expression of spirituality.

The archaeologist Ian Hodder, after more than a decade's excavation, sees Catalhöyük as laden with sacred symbols.[2] Yet he rejects the famous claim of Catalhöyük's discoverer that the site was a virtual shrine to goddess worship. Years before, James Mellaart had identified what he thought to be a Great Goddess cult centered around a female deity who controlled everything from success in farming to high fertility. In four "digging seasons" at Catalhöyük, he found dozens of voluptuous female figures that he interpreted as goddess statues, and buildings he thought were shrines for worshiping a goddess.[3] Welcome evidence to those who wished devoutly for a period of female power in prehistory, the idea of this cult became a beloved touchstone for feminists.

True enough, Catalhöyük teems with images of various kinds. In the 2004 field season alone, seventy-two figurines or clay objects, or fragments of these, were unearthed from the soil,[4] adding to hundreds more from past excavations. Elaborate murals were painted on plaster by Catalhöyük farmers who, says Hodder, were "plaster freaks." There's no quarreling with Mellaart that spirituality is bound up with at least some of these figures and painted images. Vultures and leopards, men and women, all are depicted in ways that convey concern with mortality and a groping to understand mysterious forces in the universe.

But reanalysis of the images by Hodder and his students calls into question the Great Goddess idea.[5] It turns out that most of the site's statues were not of voluptuous (fat!) females; those that are appear well after the site's first settlement. In any case, there's nothing inevitable about a linkage between fat, fertility, and female power. We know now, too, that people lived, and carried out everyday activities, in some of the buildings thought early on to have been shrines.

Further insight into religious expression at Catalhöyük comes from the work of the archaeologist Mary Voigt, who has excavated sites in Turkey for two decades. Voigt strives to put the religion of "Catalhöyük in context," as she titles one of her articles.[6] The context she means is Anatolia, the part of Turkey in Asia rather than in Europe, once called Asia Minor. One of Voigt's major contributions to understanding the Catalhöyük people is her focus on change over time in the expression of religious behavior. Indeed, from an evolutionary perspective, this site is fascinating because people lived and produced material culture there for a full thousand years.

Long-term shifts related to expression of the religious imagination at Catalhöyük do involve gender, as it turns out, just not in the ways originally supposed. As Voigt points out, the goddess theme becomes dominant only fairly late in the sequence of occupation at Catalhöyük, after about 7,500 years ago. The earliest stone figurines depict both males *and* females, some in association with animals; if these are cult figures, they represent gods as well as goddesses.

Analyzing the site's figurines by category yields some unexpected insights. The small clay figures, closely resembling objects found elsewhere in ancient Anatolia, include humans with pinched heads, cylindrical bodies, and outstretched arms; seated figures with pinched-out legs; horned animals; and animals with pronounced snouts. These are of great interest because Catalhöyük people disposed of them—clusters of them, in fact—at relatively inaccessible places, leading Voigt to think they were used within individual households as "vehicles of magic" in ritual.

Just as I have been using the terms "religion" and "spirituality" interchangeably, so does Voigt use the term "magic" to indicate one type of religious behavior. Indeed, some of these figures were fractured in ways that indicate deliberate breakage, a principal indicator that they were incorporated into some form of magic ritual.

A group of larger figures, made of clay, were found in a single building and dated to around 7,500 years ago. The biggest of these is "a monumentally fat woman seated in a chair; between her feet protrudes a small human head, presumably a child to whom she has just given birth ... the arms of the chair are formed by two large standing cats, whose tails curve up over her shoulders; her hands rest on their heads, showing her dominance over these fierce animals."[7] Here, for Mellaart, was the Great Goddess. This famous woman is now on display in an Ankara museum and is available, in replica form, for tourists to cart home from the shops. Voigt agrees that this figure represents a deity, a goddess of some sort. Yet other large human figures, some with marks of deliberate breakage, occur at the site in earlier levels. Placed near to dangerous-looking leopards, they include both males and females, and *both* sexes are portrayed in positions of control over fierce animals. A focus on both sexes in art is not unique to Catalhöyük; at sites like Nevalli Cori in the same region at the same general time period, the same pattern is found.

What is most striking about Catalhöyük is that its people do seem to become more preoccupied with the female rather than the male form as time passes and agriculture becomes more important. The theme of fat females, pregnancy, and sexuality, depicted in clay, really takes off in the later occupation of the village. Even now, though, males show up in wall paintings, so it's not that the male figure disappears altogether from the art at Catalhöyük. But the female figures are clearly meant to be deities in a way that the male figures are not.

As Voigt notes, it is simply too tenuous to leap from a focus on fat women to a fertility cult, or to a scenario in which an entire society is caught up in worship of a single Great Goddess. Still, the fat-female theme must have mattered to the worshipers at Catalhöyük; Voigt thinks the fatness-and-pregnancy theme might reflect a concern with abundance of food. This shift in artistic-spiritual expression closely tracks, in time, a shift toward heavier reliance on cultivated compared to wild foods.

Whatever the explanation for the images at Catalhöyük, it's clear that some, at least, are spiritual in nature. Our ability to see the spiritual at Catalhöyük almost certainly reflects more than mere good luck, or new and improved techniques of archaeological recovery. I am convinced that it reflects a genuine deepening of artistic-spiritual expression since the time of the European cave painters, or even since the time of the Göbekli Tepe people.

The pace of cultural change is now rapid. At Catalhöyük, farming and settled living are twinned, but the Göbekli Tepe hunter-gatherers tell us that agriculture and sedentism did not always co-occur. Indeed, from years of digging and data analysis, archaeologists are coming to grips with a conclusion that's pretty stunning: it wasn't farming that allowed people to settle down. People became sedentary *before* they began to farm.

The usual disclaimers apply: the sedentism-first, farming-second conclusion is not accepted without debate. Nonetheless, the latest evidence suggests that the hunting-and-gathering way of life, the on-the-move nomadism that had dominated the globe throughout human evolution, began to give way about 14,000 years ago. People no longer continuously roamed the land in pursuit of game, and no longer inhabited caves or huts on a seasonal basis. The dawn of our cities was at hand; people began to settle in permanent communities and, before long, to work the land surrounding those settlements.

Although sedentism and agriculture are best decoupled in our thinking, from an evolutionary perspective they originated at points extremely close in time. And with these changes, much about the expression of the human religious imagination began to shift too. Food is stockpiled now, and members of society with greater wealth and power are distinguished from others. Belongingness, stronger than ever before, incorporates a new element of rank and status. Public buildings are graced with symbols and filled with the sights and sounds of ritual practice. Clusters of permanent settlements dot a wide region, and people travel to specific locations to worship.

A full exploration of religious behavior in post–Ice Age prehistory would require many volumes of scholarship. From even a cursory look, it's clear that, in the millennia following the symbolic rituals at Sunghir and Lascaux, religious expression rooted in belongingness took off in the prehistoric world. The glories of ancient Mesopotamia and Egypt are, in a very real way, glories of ancient religion. Cultures around the world began to express the spiritual in complex and varied ways.

Just before I sent this book off to press, a new volume by Karen Armstrong was published. Curious about her analysis of patterns in early religious behavior, I read *The Great Transformation* immediately. An anecdote hidden in its midst holds, I believe, vital clues to understanding Armstrong's subject, the Axial Age (900–200 BCE):

A Chinese man named Zhuangzi (b. 370 BCE–d. 311 BCE) withdrew from public life and became a hermit. Entering a game park one day, he took aim at a magpie. The magpie, being wholly preoccupied with eyeing a cicada, did not notice Zhuangzi. Neither the cicada nor a nearby preying mantis noticed the magpie. The magpie "swept down on its prey in high excitement and gobbled them both up." A feeling of compassion welled up in Zhuangzi: he realized the creatures he had observed were fated to a sequence of mutual destruction, however unwilled. Here was the essence of life. Lost in reflection, Zhuangzi himself did not notice that he was trespassing in the park, and he was chased away by a gamekeeper.

Following these events, Zhuangzi felt depressed for months. Nothing in his life so far had prepared him for the new thoughts entering his mind: Life is about endless transformation; death should not be feared. Giving himself up to "the natural rhythm of the cosmos," Zhuangzi began to experience an "exhilarating freedom." When his wife died, a visitor who came to pay his condolences was alarmed to discover Zhuangzi "sitting cross-legged, singing rowdily, and bashing a battered old tub— flagrantly violating the dignified ceremonies of the mourning period." Zhuangzi explained that he had "cast his mind back to the time before [his wife] was born, when she had simply been part of the endlessly churning *qi*, the raw material of the universe. One day there had been a wonderful change: the *qi* had mingled together in a new way, and suddenly, there was his dear wife! Now she was dead and had simply gone through another alteration . . . If he wept and complained, he would be completely at odds with the Way things really were."[8]

This story is effective on two levels. By bringing alive for her readers a man who had lived 2,400 years ago in a distant culture, Armstrong shows us the possibility of a broad empathy. And in its focus on the redemptive nature of compassion, the role of human response to suffering in the universe, and openness to revising what we know as we live, Zhuangzi's story conveys something important about the Axial Age.

During this era, the world's religions changed dramatically in character. At the start of this period, four great civilizations were gripped by upheaval and violence. Chapter by chapter, Armstrong traces the gradual shifts in religious expression in China, Greece, India, and Israel. In each, a guiding principle was the Golden Rule.

In the first version known to history, devised by the Chinese sage Kong Qui (551–479), the Golden Rule was expressed this way: Never do unto others what you would not like them to do unto you. We know Kong Qui as Confucius, and in articulating this vision, Confucius became "one of the first people to make it crystal clear that holiness was inseparable from altruism. . . . The Way was nothing but a dedicated, ceaseless effort to nourish the holiness of others, who in return would bring out the sanctity inherent in you."[9] Armstrong destroys any notion that the Golden Rule invariably equates with an injunction to love thy neighbor. In China and the Middle East, religious thinkers were indifferent to love in our modern sense, but created utilitarian ethical visions centered around justice or practical, helpful behavior. The Greek Golden Rule emerged onstage when the dramatic chorus "issued a directive to the audience, instructing them to feel compassion"[10] for even those characters who had committed unspeakable acts.

The Golden Rule's simple wisdom became a legacy of the Axial Age. Challenged by a pagan to teach the entire Torah while standing on one leg, the Rabbi Hillel (80 BCE–30 CE) reportedly replied: "What is hateful to yourself, do not to your fellow man. That is the whole of the Torah and the remainder is but commentary. Go learn it."[11] Jesus and Muhammad exhorted their followers in similar terms to act with compassion to all people.

The Axial Age's emphasis on empathy rested on a new, and brutal, type of human awareness: Life is bound up with suffering. This uncomfortable truth must not be denied, Axial sages insisted. Further, people must strive for *kenosis*, that is, they must empty themselves of pride or ego; only then can transcendence be reached. These points are explained elegantly in *The Great Transformation*, and in its pages readers will find what has become Armstrong's trademark argument: A conflation of religion and belief is seriously misguided. "If the Buddha or Confucius had been asked whether he believed in God," she writes, "he would probably have winced slightly and explained—with great courtesy—that this was not an appropriate question."[12] Right practice, not belief, leads to transcendence.

Armstrong concludes her book with a look to the future. The lessons of the Axial Age are, she says, that people *can* reject violence and *can* effect change through compassionate action: "If religion is to bring light to our broken world, we need . . . to go in search of the lost heart, the spirit of compassion that lies at the core of all our traditions."[13]

But it's prehistory to which this book is devoted. Now that we've completed our time-line journey, let's revisit the belongingness concept and explore it in new ways. We'll return to the nonhuman world to do so.

ANIMAL ROOTS OF BELONGINGNESS

For twelve years, a deep-sea whale wandered the north Pacific, tracked by scientists at Woods Hole Oceanographic Institute.[14] Traveling all on its own, the whale roamed from the waters off California north to the Aleutians. Using deep-sea microphones borrowed from the U.S. Navy, the scientists eavesdropped as the whale repeatedly called out, trying to contact another of its kind, probably a female. As he matured, his voice deepened, just as an adolescent boy's does. No response to the whale's calls was ever heard.

What species of whale this was remains unknown, but the calls heard differed from calls of blue, fin, and humpback whales swimming in the same waters. It is a mystery why this whale received no response. One guess is that some sort of biological miswiring caused his calls to be transmitted on the wrong frequency. Another possibility is that he is a hybrid, the product of a mating between two whales of different species—and thus truly unique, with no others of his kind in the world.

Whatever the explanation, the result makes for a haunting image: a highly social and smart animal, swimming up and down the Pacific Coast for well over a decade, calling into the depths of the sea for a companion who never answered. "He must be very lonely," said one marine scientist.[15]

Whales are mammals, and mammals are social by very definition. Mammal babies suckle from their mothers, who care for them for weeks, months, or years. Most mammals live in groups, and socialize widely, with companions other than their mothers and siblings. In some cases the socializing goes on year-round, and in others, it is only seasonal. It's easy to forget that mammals evolved to be highly social with their own kind, because our contact with them tends to involve the domesticated variety. Independent, eccentric cats, and dogs bonded more fiercely with their human companions than with other dogs, are not typical mammals.

Even fish can be more socially oriented than most people might imagine. Piranhas are the poster species for fish aggression. Think "piranha" and what comes to mind but snapping jaws (the fish's) and lost digits (yours if you get too close)? As it turns out, piranhas have their own worries just like everyone else. They can be prey as well as predator; when they are under threat of predation, their breathing rate shoots up, a sure sign of stress. The rate returns to normal levels faster among piranhas in larger groups than among those in smaller ones. This is the really interesting part: these fish find comfort in numbers.[16] Granted, this form of sociality is not highly advanced and is probably not based on deep emotions; there's no evidence to say that the piranhas were attached to specific other piranhas. Still, it's an interesting and ancient underpinning to animal sociality—and in a fish.

I'm no ornithologist, but the little reading I have done on bird behavior makes it clear that birds do have emotional responses. Berndt Heinrich's *The Geese of Beaver Bog* describes the intense loyalty between a male Canada goose and his chosen female partner. Though it would be easy to write off such bonding as instinctual (coded in goose genes), Heinrich's careful descriptions take us instead into a world where the geese each possess a unique history and set of relationships with other individual geese.

One of my favorite animal anecdotes supports Heinrich's point about geese relationships very well; I borrow it from the newsletter of an animal rescue society located near Santa Cruz, California:

> Last spring a female goose was brought to us from Pinto Lake with a fishhook in her leg. We took her to the vet; we had to cut it out since it was so deeply imbedded. Then we brought her to our intake facility and gave her antibiotics. When she healed sufficiently, we let her swim in a small pool.
>
> About ten days later, a man called us about a goose walking down Freedom Blvd. [a busy road running from the lake towards Santa Cruz]. Fearing it would get run over, he brought it to us. When he arrived, he was carrying the (male) goose in his arms while the female goose that we rescued earlier swam in her pool.
>
> As soon as the male goose saw the female, he started to honk, jumped out of the man's arms and ran toward the pool. The female began to flap her wings and thrash around in the water. Then, the male jumped in with her, put his neck over her and angrily hissed at us. From then on, if anyone came near her, he would put his neck over her protectively, hissing a warning to stay away.
>
> We believe he was her mate and had been looking for her. It was amazing that he was able to walk all the way to Freedom Blvd. from Pinto Lake!
>
> We found a safe home for both of them on someone's property with a large pond, where they're living happily ever after.

Let's play hardcore skeptics for a minute. How can anyone know that the male was seeking a specific female, his remembered mate? Isn't the only valid test of this notion to put *two* female geese in *two* nearby pools, then bring in the male to see whether he flies to, and protects, only the injured female? But even this result might not be proof enough. The male might have acted on a general impulse to protect a vulnerable female of his species, in the absence of a specific bond. All I can say to skeptics is, Read *Geese of Beaver Bog!* Heinrich is very convincing on the enduring attachment between specific males and specific females.

My overarching point is that mammals have no monopoly on being social, or on feeling emotions. All species are well adapted to their own habitat and social needs (or lack of them); at a basic level, belongingness may be widely selected for in evolution. Nevertheless, my money is on an evolutionary continuum: more than fish, birds, reptiles, or insects, mammals are likely to express emotional connections with each other as social events unfold between relatives and companions. Among mammals, the primates (and perhaps dolphins and elephants) are likely to show the most elaborated of these.

Long-term study of the baboons at Amboseli National Park, where I did my feeding study years ago, reveals an evolutionary advantage to monkey sociality. Female baboons (and other monkey females as well) express long-lasting bonds with social partners through grooming, mutual support during conflicts, and a relaxed hanging out together. The primatologist Joan Silk and her colleagues drew on sixteen years of Amboseli baboon data to show that the females who were more socially integrated into their groups were more likely than their less social counterparts to rear infants successfully.[17]

Amassing enough data on long-lived primates to reach a conclusion like this is rare: this study is renowned in monkey-watching circles. The researchers avoid interpreting their data in terms of baboon popularity or baboon loneliness, but they do note that feeling lonely can affect health and life span in humans. Whatever the mechanism, the result is that for monkeys, socializing can lead to success in reproduction. In evolutionary terms, this practically guarantees that a talent for socializing should be selected for, and passed on from generation to generation.

In using animals to shed light on humans, more useful than

whales, piranhas, geese, or even monkeys are the African great apes. Chimpanzees, bonobos, and gorillas inherit the basic mammalian social legacy, plus of course the monkey social legacy, then add something more. We have seen the special way in which these apes co-create meaning by responding to each other contingently when they interact. A mother chimpanzee closely observes her daughter's attempts to crack open a tough nut with a stone. She intervenes to provide guidance at the right moment, enabling the daughter to change her technique just enough to be successful. Two excited and potentially aggressive male chimpanzee brothers gesture at a third, quieter male sitting alone: an attempt to recruit an ally in an ongoing rivalry. A mother gorilla moves quietly to stand near her two juvenile sons when their play, suddenly becoming too rough, causes the younger son to utter distress vocalizations. At her approach, the pair breaks off playing, and the younger son runs to the mother and embraces her.

Because the apes represent our evolutionary platform, our ancient primate legacy is to *be transformed moment by moment through the process of being social and emotional with others.* Hominid belongingness, as reviewed in the previous three chapters, expanded from this base. The hominid brain grew from its initial apelike size and, more important, increased in complexity, with the ability to think in symbols underwritten by changes in how the brain is organized and by increases in neural interconnectivity. We'll see later in this chapter how these cognitive changes may have been driven by increasing emotional engagement in hominid social groups and families. For now, let's look more closely at belongingness in our own species. What does belongingness mean in our lives today?

I aim to answer this question in a few different ways, each lodged in the evolutionary perspective. As John Donne famously wrote four centuries ago, "No man is an island." Humans' essential need to be near others and to connect with them emotionally is nowadays subject to intense study by scientists who focus on human belongingness or its evolution. In the remainder of this chapter, I discuss some of these studies, some more concerned with data collection and others more

with theorizing; all offer a path to understanding ways in which belongingness today is emotional, adaptive, and connected to the religious imagination.

RELIGION AND BELONGINGNESS

Belongingness involves frequent, positive interaction within ongoing
relationships. Without it, humans suffer: "People who lack belongingness suffer higher levels of mental and physical illness and are relatively
highly prone to a broad range of behavioral problems, ranging from
traffic accidents to criminality to suicide."[18] To put the matter more
positively, emotional connections not only feel good, they also help ensure good health and longevity. Psychologists think in terms of a need
for belongingness in humans, not just a want; this view makes perfect
sense; since we evolved from apelike ancestors who lived in socio-
emotional connection to others in their group.

Quality of belongingness trumps quantity every time. People
rarely try to accumulate more and more close relationships, piling one
atop another to seed our lives with intimacy at every turn. Even people
firmly rooted in the center of a broad social network, or a wide circle of
friends, tend to focus on only a few keenly rewarding relationships.

Sometimes one intensely felt relationship is enough, at least for a
while. How many of us have had a close friend, intoxicated by a new
love, become less available emotionally to us? Even if we feel genuinely
happy for our friend, and intellectually accept the reason for her with-
drawal, it may still hurt in our hearts. Psychologists confirm that our
feeling of neglect in this situation is rooted in reality; new love does
seem to "supplant" other ties "and satisfy the belongingness need previously satisfied by the other friendships."[19]

Literature can help us out here. In *Madame Bovary*, Gustave
Flaubert masterfully describes his characters' emotions as new love in-
toxicates them. When Leon worries that he loves Emma more passionately than Emma loves him, or when the two are forced to be apart,
bliss shades into anguish. Note the element of yearning imagination,

and the richness of the inner life surrounding the beloved, as Leon talks to Emma: "How I've suffered! Many a time I would take off, plod along by the river, to addle my brain with the noise of the crowd, yet quite powerless to banish the obsession that was hounding me. . . . Often, I used to write letters to you and tear them up. . . . I thought that I saw you at the corner of the street." His words transform Emma: "It was just like the sky, when a puff of wind sweeps away the clouds. The burden of sadness that dimmed her blue eyes seemed to be lifted; her whole face was shining."[20]

Flaubert's words resonate with us. Most of us have lived our own love-saturated moments (though, happily, without at last resorting to arsenic as Emma does!) Such intensity is felt even outside of romantic love. When those of us who are parents interact with our older children, we remember the besotted feeling of falling in love during the baby's magical first year. Attending the funeral of an acquaintance, we may find ourselves in tears. We grieve not only for the dead person's friends and family but also for our own remembered losses as well. The intensity at the core of our deepest relating is familiar to us all.

New research shows that when someone we love feels physical pain, our brain responds as if we felt it.[21] Researchers may have known people who were especially eager to advance the cause of science or perhaps they just rounded up college students who needed a few extra bucks: in any case, they managed to locate thirty-two people, sixteen heterosexual couples, willing to receive one-second-long electric shocks. All the women received the shocks while inside an MRI machine.

Shocks were delivered either to the man's hand or to the woman's. Cues visible to the woman informed her whether the forthcoming shock was headed for her hand or her partner's, and also whether the shock would be weak or sharp in intensity. The researchers then monitored the response of the woman's brain. Naturally, when she herself was shocked, her brain's pain centers went into action. The significant finding is that many of these same pain areas lit up when her partner got the shock, even though she herself felt nothing. Nothing physically, that is; she did feel the shock emotionally. When we love someone, we really do feel their pain; we imagine it and we empathize with them. (It would be good to know whether the men's brains feel their partners' pain as much as the women's brains did; perhaps that study is in the pipeline.)

Let's return to the nature of primary relationships. Even when intensely felt, they are rarely insular for long; they often lead, in ever-widening circles like ripples in a pond, to other emotional connections. Because we are creatures made to relate, our behaviors may be swayed by what our social partners do. Relying once again on the magic of MRI brain-scanning, scientists uncovered a fascinating link between people's problem-solving behaviors and their social involvement with others.[22] In countries like the United States, where independence of thought is highly valued, the study's take-home message may be received soberly: when an individual's personal judgment is at odds with that of others in a group, he or she tends to conform to the group position.

Thirty-two volunteers were tested on a mental-rotation task. These people were asked to decide whether certain three-dimensional objects, when differently rotated, were the same or different. This task is familiar to those of us who remember the line drawings on high-school aptitude tests: figure A must be compared with rotated figures B, C, and D, only one of which is identical to A. In the study, the volunteers were informed of the answer selected by four other participants. These other participants, assumed by the volunteers to be their peers, were in reality actors in cahoots with the researchers.

Before actual testing began, the volunteers and the actors informally interacted. This phase was engineered to enhance the social bonding between the two groups. During the testing phase, actors intentionally offered false answers half the time. Previous research had led the scientists to expect that at least some of the volunteers would go along with the actors' incorrect choices. And their knowledge of the brain led the scientists to make some interesting predictions, as follows.

When volunteers echo the actors' incorrect answers, and thus socially conform, the MRI should show activity in one of two brain areas, depending on the reason for the conformity. When participants make a strategic decision to go along with the group (despite knowing that the group is wrong), changes should be observable in the prefrontal cortex. By contrast, when they choose the incorrect answer because they perceive the task itself differently as a result of knowing what others in the group had answered (and do *not* know that the group is in error), changes in the occipital-parietal area should occur. In order to test these predictions, volunteers had to be interviewed about their decision-making processes after their brains were scanned.

The study's results appear in an article crammed full of methodological details and varying potential explanations for the results. But let's cut to the bottom line: people did conform to the group, and at a startlingly high rate: 41 percent of the time, far more than when computers instead of live actors were selectors of wrong answers. Clearly, the volunteers were swayed by what their human companions had to say, although they had just met these people a short while before!

The most exciting finding is this: people conformed not because they assessed that the clever thing to do would be to join with the others, as a calculated social strategy. No, people's very perceptions of the task shifted simply because they learned how their peers (in reality, the actors) had answered. The MRI results tell this story quite clearly.

Of course, not everyone conformed. What happened to the nonconformists is fascinating: their brains got emotional. That is, the brain activity of these independent thinkers reflected emotional involvement. This is a red-letter finding, because it shows that such independence is linked to an "emotional load associated with standing up for one's belief," as the scientists put it. On two counts, then, this study has wide-ranging implications. Social involvement may alter a person's perception of the world, and it may be emotionally costly for humans to go against the crowd.

Before relating these findings to religion, I want to consider their link to everyday life—life outside the psychology lab, where we interact not with researchers and actors but with people we truly care about.

Let's say you passionately follow a certain sports team. From the world of American baseball my husband would nominate the Boston Red Sox; many people from Brazil to the UK would choose a soccer team. Chances are you will have acquired this passion from, and share it with, someone you care about. This person may be a grandparent, parent, aunt or uncle, older brother or sister, or friend from school days. Serious sports enthusiasts don't merely tune in to the action on radio or TV while at home alone. They yell enthusiastically together as they watch games with people who care just as much; together they avidly discuss the fates of individual players and track their seasonal statistics, whether with delight or disgust.

Or maybe your thing is music. As a totally random example, consider the level of dedication some fans express for Bruce Springsteen and the E Street Band.[23] Constantly in touch via the Internet, these fans coordinate plans to crisscross the United States and Europe in order to be in concert with Springsteen and with each other. (I'm a minor-league groupie by comparison, having attended Springsteen shows in only five American states.) Not content to listen at home to their CDs or their iPod downloads, these folks pool ideas and responses as they analyze lyrics, trade CDs, and bring their children to the concerts as a rite of passage.

Belongingness experienced with thousands of others at a Springsteen concert is without question a high. On the right night—where Springsteen is concerned, this is nearly every night—the band and the audience join together to co-create a visible and audible *emotional* response to the music. Sitting under the stars in an open field, waiting for just that moment, the moment you know is coming, when the sax pierces the night and underscores the lyrics of longing and desire, is transformative. It brings you into contact with a stream of energy that flows inside and outside your body all at once. Some concertgoers express this energy as spiritual in nature.[24] This phenomenon has been noticed by the theologian Kate McCarthy: "People [who attend rock concerts] talk about feeling connected to something larger than them-

selves, or about having feelings of ecstasy," she says. "So I've always wondered how what's going on parallels religious experience, and how we
relate the two vocabularies."[25]

In sports, music, or any other human endeavor, it comes naturally
to our species to share what we believe and feel with those we spend
time with and those we love. This is part of belongingness, and it
means that we are affected emotionally in return by what others believe
and feel. No literal aping need be implied here; we may reject others'
choices as much as we embrace them (just ask any parent of a
teenager). The key pattern is that we tend to be emotionally affected by
our social partners in some way, a pattern that applies readily to the expression of the religious imagination.

The pull of a social network, in fact, is the single strongest factor
in why people convert to a new religion or join an established religious
group. People become attached to those who already belong, and are
drawn in. This social pull far exceeds the lure of doctrine or ideology,
a finding that only long-term research like sociologist Rodney Stark's
could reveal: "When people retrospectively describe their conversions,
they tend to put the stress on theology. . . . [But] we [researchers]
could remember when most of them regarded the religious beliefs of
their new set of friends as quite odd."[26]

This finding holds for religious conversion in the present day and
historically, in our society and in others. The roots of Islam thousands
of years ago and the trajectory of Sun Myung Moon's Unification
Church in twentieth-century America are both social in nature—and,
as we have seen, this means *emotional* in nature. In *The Rise of Christianity*, Stark focuses on the blossoming of Christianity from its roots in a
pagan world. Early Christianity involved conversions along social-
network lines, and, Stark writes, brought about genuine social transformation as it spread to a pagan, Roman world teeming with social
ills.

"To cities filled with newcomers and strangers," Stark writes,
"Christianity offered an immediate basis for attachments. To cities
filled with orphans and widows, Christianity provided a new and ex-

panded sense of family. To cities torn by violent ethnic strife, Christianity offered a new basis for social solidarity."[27] Innovative ways of social relating, at least for this time and this place, emerged as Christianity found its footing.

The views of a person's social group, then, may affect a person's perceptions about religious matters no less than about the rotation of figures in a problem-solving task. Perhaps researchers will design an MRI study to test this idea in the lab. Yet the expression of the religious imagination is not on a par with social interactions in a lab setting, or even with matters of everyday social life. With the religious imagination, we move away from an earthly belongingness and into the realm of relating to a partner who is neither visible nor knowable (and with far greater emotional depth and hope for achieving transformation than is experienced in relating to a fantasy creature such as Santa Claus). This is what Pascal Boyer means when he says that religion involves beliefs—whether related to the ghost banquets of the Fang in Africa or the Holy Trinity of Catholics everywhere—that are counterintuitive or extraordinary in certain patterned ways.

Yet a focus on belongingness takes us beyond Boyer (as we'll see in the next chapter). It takes us beyond belief as well. As we have seen, the human tendency to engage with God, gods, and spirits emerges from a long evolutionary history of emotional transformation between individuals and within groups. Conversions along social networks could not occur before a religious imagination emerged in the first place. Stark claims that the ideas of a loving God and of the rightness of expressing love of God through love of others were inventions of Judeo-Christian thought. For pagans, the relationship revolved around self-interest: if you treat the gods appropriately, the gods in return will help you, and ease your way in life.

Within an evolutionary framework, a change that has taken place within the historical span of Judeo-Christian ideology is quite recent; indeed, the social science literature commonly describes changes in religious expression that occurred just yesterday, on our time line. Tanya M. Luhrmann suggests, for instance, that an intense bodily experience of God, particularly through trance, has made worship within U.S. Christianity a "markedly different" phenomenon over just the past three or four decades: "What is striking about U.S. religion since the 1960s is that it not only emphasizes bodily phenomena but also uses those experiences to create remarkably intimate relationships with God."[28]

Most social scientists think historically, not evolutionarily. Stark's goal, for instance, is to contrast early Christianity with the paganism of the day. Whether he is right or not about specific emotional innovations within Judeo-Christian thought, it's worrisome that he makes bold claims like this one: "What was new was the notion that more than self-interested exchange relations were possible between humans and the supernatural."[29]

An evolutionary perspective reaches a different conclusion. Since the time of early *Homo sapiens*, maybe even since Neandertals, people have shared emotion—better yet, people have *created* emotion—as they participate in religious ritual. They *feel* as they relate to God, gods, or spirits through active engagement in ritual; they experience trance and other altered states that bring them closer to sacred forces. This is more than "self-interested exchange relations."

As before, it may help to seek analogy with traditional societies. African Pygmies saw the forest they inhabited as the living expression of the central deity in their lives. Through ritual carried out in the forest, the Pygmies expressed not only their needs but also their affection for this deity. Each person in the group, writes the anthropologist Colin Turnbull, "enters a relationship with the deity in which he or she assumes that the forest will naturally reciprocate [their] offerings just as parents respond to their children by caring for their needs."[30]

The emotional transformation wrapped up in such intense relating with God, god, or spirits is a cross-cultural pattern with deep roots. A pattern is defined by features that recur and make up a stable invariant core; but it is also characterized by some degree of variability in how that core is expressed under specific local conditions. Exceptions are central in helping us see the pattern in the first place, because patterns take on coherence to the degree that they stand out from what else is going on in the universe.

When one is seeking exceptions to expressions of the religious imagination, Buddhism comes in very handy. As we have seen, Buddhists do not believe in one supreme deity, yet Buddhist gods and goddesses surely exist, and emotional transformation through compassionate action is paramount.

At Buddhism's center are four noble truths. Life is suffering (the first truth), because attachment to things, outcomes, and desires is part of being human (the second truth). Suffering can, however, be overcome (the third truth) via a path of right living, illuminated by certain virtues that should guide our lives (the fourth truth). The chief virtues are helping those in need; feeling compassion for the suffering of others; feeling the joy of others; and accepting what is.

The philosophy underlying Buddhism is complex. Buddhism differs from Judaism and Christianity in its disavowal of a supreme entity and in the degree to which it counsels a rejection of attachment. Yet it shares with these religions both a commitment to compassion and emotional relating with others, and the hope of being transformed through those actions.

From religion to religion, the details of ritual practice and emotional transformation vary. In some cases, direct emotional connection with a supreme other is the central desire; in other cases, there is a more diffuse connection with sacred forces. As with Boyer's Fang and their ghosts, the emotional connection may be based as much on anxiety and fear as on nurturance and love. Further, as Karen Armstrong has explained, during certain historical periods, prophets and sages insisted that an intimate knowledge of God was impossible. The ultimate was thought to be "ineffable, indescribable, and incomprehensible—and yet something that human beings *could* experience, though not by reason."[31] This point is important, because it would be a serious oversimplification for me to imply that throughout the ages, people felt they *knew*, or *could know*, God (or gods or spirits) in an intimate way. I believe that the desire for emotional connection with the sacred is fundamental to our species, but this desire has been expressed in a myriad of ways throughout the centuries. Ineffability does not stand in contradiction to belongingness.

The religious imagination thrives on the human yearning to enter into emotional experience with some force vaster than ourselves. This pattern, then—in its essence rather than in its details—stretches back far into our prehistory. For millions of years, human ancestors sought belongingness within their social groups; as they continued to evolve physically, behaviorally, culturally, and spiritually, humans began to seek an emotional connection with God, gods, or spirits. What happened was gradual rather than a spiritual "big bang." The human religious imagination developed in ever widening circles of engagement from immediate social companions, to members of a larger group, then across groups, and, eventually, to a wholly other dimension, the realm of sacred beings. Australopithecines and early *Homo* engaged, almost certainly, in complex interactions with their social partners that were based on remembered social histories and emotional nuances, just as is the case with the African apes. Later, as these tendencies expanded in scope, they twinned with greatly enhanced abilities in the "big four": empathy, co-constructed meaning-making, rule-following, and imagination.

Neandertals began to bury their dead via symbol-laden ritual, on occasion via ceremonies that look to archaeologists like funerals with communal feasting. The first stirrings of jewelry-making, music, and art emerge, too, in Neandertal groups. Early modern *Homo sapiens* began to decorate their bodies, both in life and in death, in ways that announced relative status and group identity. Images these hominids painted, some with shamanic overtones, grace inaccessible and dark cave passages. Ritual, including art-centered ritual, becoming more and more elaborate over time, helped these humans make sense of the world. Perhaps, given their intense emotional relating, big brains, and changing relationship to the environment through deepening symbolic ritual, they imagined an afterlife or spirit world. The weight of the evidence strongly suggests that *Homo sapiens* began to engage in *some* way with sacred worlds, thus further transforming their lives as they continued to evolve. By the time of sites such as Göbekli Tepe and Catalhöyük in Turkey, the religious imagination was codified in buildings where ritual worship was enacted.

TRANSFORMATION IN PAST AND PRESENT

At its deepest roots, religion is transforming *because* it is active and emotional. As one's very depths are stirred through connection with other people and other forces, especially during ritual, change occurs. Neither the individual nor the group stays the same during or after participation in religious ritual or other collective expression of emotion. This transformative process was understood by Emile Durkheim, the great French sociologist and philosopher. He objected to a purely cognitive understanding of religion:

> Most often, the theorists who have endeavoured to express religion in rational terms have seen it, above all, as a system of ideas that correspond to a definite object. This object has been conceived in different ways. . . . In all cases, it was ideas and beliefs that were considered the essential element of religion. . . .
>
> But believers, men who live the religious life and sense its substance directly, object that this way of seeing does not correspond to their daily experience. They feel, in fact, that the true function of religion is not to make us think, to enrich our knowledge, to add thoughts from another source and of another kind to the thoughts we owe to science, but to make us act, to make us live. The worshipper who has communed with his god is not only a man who sees new truths that the unbeliever does not know; he is a man who is *capable* of more.[32]

Here is an historical foundation for a definition of religion as grounded in action. Yet Durkheim goes even further: "The collective ideal that religion expresses, then, is . . . due to . . . the school of collective life that the individual has learned to idealize. . . . It is society that, by bringing him into its sphere of influence, has infected him with the need to raise himself above the world of experience and has, at the same time, provided him with the means of conceiving of another. Society has constructed this new world by constructing itself, because it is society that this new world expresses."[33]

What can Durkheim mean by asserting that society constructs itself? One way to understand this notion better is to hurtle from the sociological past to the psychological present. Two sets of ideas that address the transformative power of human emotion are quite helpful here. When wedded to my own evolutionary account centered on empathy, meaning-making, rule-following, and imagination, these ideas strengthen the case for belongingness as the engine for the origins of religion in prehistoric times.

ART AND INTIMACY

The Yekuana people of Venezuela and Brazil live (or, at least, they once did) by means of hunting, gathering, and horticulture. As in many such traditional societies, ritual is woven into everyday acts in this society, sometimes quite literally. Ritual, says the arts scholar Ellen Dissanayake, permeates every Yekuana activity, no matter how mundane: "All actions communicate the same essential messages and meanings: 'To tell a story, therefore, was to weave a basket, just as it was to make a canoe . . . to build a house, to clear a garden, to give birth, to die.' "[34]

The Yekuana creation myth, and thus the group's way of being, is grounded in a strict dichotomy between the wild and dangerous, on the one hand, and the tamed and safe, on the other. This perspective on the world is not taught explicitly to youngsters, but simply exists in routine acts, and comes to be known through these acts. Manioc (*tapa*, in the local language), the single most important Yekuana food, is toxic in its natural state, for example. Before eating manioc people must process it, and transform it through cultural practice from a wild state to a safe state.

The theme of transformation plays out over and over again in Yekuana daily life. Animals and people, and even vegetables, are understood to have spirits that live in a sort of parallel world. These spirits are potentially troublesome but also potentially controllable, just like *tapa*. In sum, all acts of life, from eating to dealing with the spirit (sacred) world, are about transforming the wild and dangerous to the cultural and safe.

Building upon the Yekuana example and others from traditional societies, Dissanayake herself produces a beautifully realized interdisciplinary framework for understanding how ideas of art and religion are embodied in, indeed are created in, everyday life. The tapestry she has woven on that framework is her book *Art and Intimacy*.

In bringing together mother-infant mutuality; creativity through rhythm and mode; and the evolution of art and religion, Dissanayake speaks the language of meaning-making at multiple levels. "It is not surprising," she writes, "that societies all over the world have developed these nodes of culture that we call ceremonies or rituals, which do for their members what mothers naturally do for babies: engage their interest, involve them in a shared rhythmic pulse, and thereby install feelings of closeness and communion."[35]

Joy, for mothers and babies, is created by shared communication and shared movement. Recognizing this is amazingly easy once you look beyond the words uttered (or syllables babbled), because infants are very physical. The tricky thing is to see the rhythm and mode in highly verbal human interaction as well. As always, let's operationalize. For Dissanayake, "rhythm" means the patterned way in which an experience unfolds in time. By contrast, "mode" refers to the qualities of that unfolding, "its sense of swiftness, solidity, opening, closing, speed, forcefulness, fullness, barrenness, lightness . . ."

These two terms resonate with me. In my own research, I try to notice and then describe, patterned differences in the quality of body

Mother and baby engaging each other through shared communication.   *Corbis*

movement and gesture in apes. To say "The silverback gorilla ran across the cage" is to neglect rhythm and mode. Did the male make a showy display run while tensing his muscles and holding his limbs rigid? Or did he gambol in a loose-limbed and playful way? Do the male's interactions with social partners play out differently according to the variations in the rhythms and modes of his behaviors?

Investigating rhythm and mode takes us a good distance from art and religion as they are typically understood. That's fine with Dissanayake, for she wants to move both art and religion out of the realm of symbols (a mental world) and into that of movement (an emotional world). Art is about "the impetus to elaborate." Art isn't art *unless* it is elaborated, made special in some effortful way through love's labor. More than just using symbols to represent some aspect of the world, art is about doing in the world.

Think back to a visit you made to a museum with paintings on its walls, sculpture in its garden, and a large open space devoted to performances in dance or music. What, for you, was the art the museum was meant to show off? Was it the pink whirl on the wall, Degas's ballerina? Or the fluid lines of mother embracing child, in Allan Houser's motherhood sculpture series? Degas and Houser made art, of course. But art is in the process as well as the product: Degas stroking pink on to the canvas, Houser shaping two linked forms from raw material as he moves around in three-dimensional space. It's the same with the dance ensemble or the chamber music trio: the art is in the coordination of the limbs as dancers move across the stage, in the bow's being drawn over strings as the violinist plays in harmony with the soft striking of the piano keys.

The movement *is* the art. The art emerges from creative rhythm and mode, and sometimes from that blissful sense of losing oneself in the flow of rhythm and mode. Remembering the movements of creating art, and not just the final product, is like remembering the movements of expressing one's love (whether for an adult lover or for an infant) and not just the love itself.

How shall we apply this insight to prehistory? Cave paintings are, of course, stunning examples of prehistoric art, but the notion of embodied elaboration challenges us to think more broadly. What about the process of the painting? What, Dissanayake asks, about dancing (or even just moving rhythmically), singing and chanting? These activities leave no fossilized traces, but would be overwhelmingly likely to occur in people with the cognitive and emotional capacities of Neandertals or early *Homo sapiens*.[36]

"When you boil it all down," says Dissanayake, "that is the social purpose of art: the creation of mutuality, the passage from feeling into shared meaning."[37] Religion, too, is about creating shared meaning through action in the world, and indeed, religion and the arts developed together in prehistory, enabling and enhancing each other.

Humans have evolved, in Dissanayake's framework, to seek, and to thrive on, the very mutuality we all experience as infants, just as much as they seek and thrive on nutritious food and safe shelter. The back-and-forth wide-eyed and aroused mutuality of human mothers and babies is adaptive far beyond the young years; its rhythm and mode become the "means for arousing interest, riveting joint attention, synchronizing bodily rhythms and activities, conveying messages with conviction and memorability, and ultimately indoctrinating and reinforcing right attitudes and behavior."[38]

Dissanayake's mutuality is somewhat like belongingness, but is more creative. Mutuality is not just a need, and it is not innate; rather, it is a force of life that we create in each other as we live. Why mutuality cannot be simply an innate need is made crystal-clear by engaging with another set of ideas that centers around emotional transformation.

THE FIRST IDEA

Picture a child, Beth, as she grows from a plump infant into an active toddler and then an increasingly social, language-using preschooler. Beth the toddler flies into a rage when her older brother seizes a piece of candy from her hand. She cries inconsolably when her mother goes out for an evening, leaving her with a baby-sitter. But as she enters her third and fourth years, Beth learns how to calm her emotions and how to stop herself from responding so violently. She does this via symbols. She can picture for herself the piece of chocolate that she got two days ago, when she asked for it politely (and she definitely retains the image of her impolite brother not getting one!). She can recall the warm hugs that envelop her when her mother returns from an outing (and she soothes herself by recalling that Mom always does return, every single time).

But this engagement with symbols is not lodged entirely in the memory. In fact, the source of Beth's cognitive development is emotional—it lies in emotion-based nurturing and caretaking. This is the thesis of *The First Idea*, written by my colleagues Stanley Greenspan and Stuart Shanker, and a perfect companion to *Art and Intimacy*.[39] For Greenspan and Shanker, the ability to separate perception from emotional action is the key to understanding change, whether during the lifetime of a single child or over evolutionary time.

What does it mean to separate perception from emotional action? It means being able to weigh our choices, even in the face of overwhelming rage, or fear, or joy, or desire. It means being able to think calmly before we act, using words or images to weigh the potential consequences of behaving one way or another. At times, it means using language to completely *replace* action. These abilities develop gradually, both in the lifetime of a single child, and over evolutionary time in the primate/hominid lineage: "What takes a human baby two years to learn took our human ancestors millions of years."[40]

During the first year of life, "caregivers help babies begin to learn how to transform catastrophic emotions into interactive signals."[41] This process really takes hold during a child's second year. From then on, emotional communication with a nurturing caretaker drives the development of ever-higher levels of thinking.

To understand better how this happens, let's return to Beth for a moment. Picture her at age two, confined indoors with her mother during summer vacation at the beach. It's an infinitely rainy day. Wanting badly to build sand castles at the water's edge, not to be stuck inside, Beth expresses her frustration by throwing books around the room and by hitting the family dog. Overwhelmed by catastrophic emotion, she has "acted out," quickly traversing a direct path between her perception of a situation and her emotional response to it.

The next year, the same thing happens to this meteorologically unlucky vacationing family: constant rain. Now, though, Beth "can symbolize hitting and screaming by saying 'Me mad!' "[42] How did this transformation occur? Was the change lurking within the girl's genes, awaiting her third birthday to flower into expression? Was it just a matter of time and linguistic maturation, so that expressing her feelings with words, instead of by throwing and hitting, was an inevitable outcome?

Neither genes nor brain changes alone can account for the change in Beth. As *The First Idea* explains, reciprocal, co-regulated emotional interaction drives this kind of change. "Reciprocal" simply means that both parties, child and caregiver, actively participate, and "co-regulated" means that each party shapes what the other does as the action unfolds. Here, Greenspan and Shanker are describing a process similar to the one by which African apes co-construct meaning across the generations, though in humans the behaviors are more complex. Specifically, the human caretaker takes the lead role in shifting the child away from her overwhelmed state via an intimate back-and-forth dance of engagement.

Picture once more our housebound little vacationer. At one point, as the rainy day drags on, Beth's muscles begin to tense, and her facial expression tightens. But before she can reach for an object to hurl, her mother intervenes:

> If the caretaker responds before the child actually bites or hits, that is, responds to the *intent*, the child is likely to respond with another *intent*. For example, the parent responds to the child's angry look with a soft soothing look of "what's the matter" and with hands out, an offer to pick her up and cuddle or feed her. The child responds with a softening of her grimace and anger and a look of expectation. The parent then responds with another gesture— reaching closer to the child—and the child now begins to break into a smile as she reaches her hands toward the parent.[43]

The child calms, and does not launch into an aggressive acting out.

What's critical here is the back-and-forth. Over time, with hundreds of such interactions, each slightly different from the last, the child is challenged to modify her response of acting-out-based-on-emotion. The child is a full partner in the nurturing context; she and her emotions are thereby transformed.

Of course, not all caretakers around the world express love and nurturing in the ways highlighted by Greenspan and Shanker, who work with babies and young children in North America. In some societies, parents prioritize instruction and protection rather than explicit expressions of love; overt concern for the baby's emotional regulation may even be frowned upon. Yet all babies and their caretakers (except in cases of abuse or neglect) participate in reciprocal, co-regulated interactions based on deep positive emotions of some sort. As always, it is the pattern that we seek.

Now we are in a position to see why it's critical to resist any automatic equation of "universal" with "innate," a trap even some savvy social scientists fall into. Belongingness and mutuality may be universal human needs, but they are not "innate" in the sense of being encoded in the genes. Because intimacy comes from belongingness, present from the first moments after birth, it is created by what we do with others.

What about change over time in hominids? As bigger brains and enhanced capacity for reciprocal co-regulated communication evolved, changes in symbolic and reflective capacity developed, too: "The underlying cause of the evolutionary changes manifest in the fossil record—morphological, behavioral, cognitive, and social—has been in part due to a series of advances in caregiving practices built on the practices of preceding generations."[44]

Thus the framework of *The First Idea* fits beautifully with a wish to see hominids as more than tool makers, hunters, and long-distance migrators. The evolutionary changes in hominid belongingness were bootstrapped by patterns of engaged emotional nurturing, patterns that entwined with the development of symbolic thinking in a beautiful feedback system.

TAKING STOCK: THE BIG PICTURE

Let's integrate the insights gleaned from Dissanayake's writing on mutuality and Greenspan and Shanker's research on co-regulated emotional communication with the belongingness framework for the origins of the religious imagination. We are born into mutuality. Throughout our lives, we are continuously transformed by mutuality. Our ape roots involve not just grasping hands, depth perception, big brains, and being social, but also empathy, co-construction of meaning, rule-following, and imagination. As our brains grew ever larger and more capable of fueling sophistication in these practices, caretaking practices changed too, to reflect greater co-regulated emotional communication.

Human ancestors began to experience mutuality in ever deeper ways. Hominids began to express themselves through self-decoration, rituals for the dead, the rhythmic modes of music and dance, and the making and appreciating of art. Some of these experiences, whether in dark caves or open-air sites, brought hominids into a different plane of relating, an imaginative plane that brought them at least to the verge of relating with God, gods, and spirits.

People today relate intensely with God, gods, and spirits. Following Karen Armstrong, we have defined religion today as bound up with compassionate action. This is fine if we are talking about ideals. Anyone with a passing knowledge of human history might with justification react cynically to a suggestion that religion equates in reality to compassionate action, either now or in the past. The Crusades are an obvious example, producing more than a million deaths during 176 long and bloody years. From our own time, we may look to the African country of Sudan, with 2 million deaths in a twenty-year, religion-driven civil war as one egregious example among too many. Violence doesn't exist only on other continents and in past eras. Our own tragic history in the United States reeks of intolerance and hatred mired in boasts of religious righteousness. The Ku Klux Klan's campaign against blacks in the American South was an avowedly Christian campaign, to take but one example. This is belongingness gone awry, belongingness that becomes about group versus group.

Yet emotional transformation was present from the first stirrings of the human religious imagination. The expansion of belongingness to a spiritual realm emerged from mutuality and from hominids' talent for constructing their lives around emotional engagement with others.

Once again we bump up against the question of whether humans discovered or created God, gods, and spirits. My answer remains agnostic. Karen Armstrong once again offers insight, this time in *A History of God*: "Ever since the prophets of Israel started to ascribe their own feelings and experiences to God, monotheists have in some sense created a God for themselves. God has rarely been seen as a self-evident fact that can be encountered like any other objective existent. Today many people seem to have lost the will to make this imaginative effort."[45]

No contradiction need exist between deep faith and imaginative faith. The seamlessness between religion and everyday life that exists in traditional societies now is based not on knowing God, gods, or spirits, but on moving through the world with a spirituality that is part of everything that is and everything that one does. Religion is of the body, the mind, the imagination, of relating and loving and caring all at once.

Yet what is at the forefront of the push to understand early religion right now? Scholars focus on genes and hard-wired brains. Why should this be so, and how useful are the genetic and neurobiological approaches?

SEVEN

Is God in the Genes?

O VER AND OVER, I have returned to the primacy of the social, the emotional, the imaginative, and the transformative
in explaining the origins of the religious imagination. Yet this cluster
gets little airtime in the current climate, when many investigators seem
fixated on "God genes" and on brains hardwired for religion. Let's take
a look at how far genetic explanations can take us in understanding the
origins of religion.

One day while holed up in my study, writing the first chapter of
this book, I heard a telltale *thunk* against my front door. The sound of
a new book arriving via UPS is a familiar one at our house. That day's
package bore an eye-catching title by Dean Hamer: *The God Gene: How
Faith Is Hardwired into Our Genes.*[1] Here is a title to rival Pascal Boyer's
brash *Religion Explained*. The idea of a single God gene certainly caught
my attention, and I was far from alone.

Soon after that day, I was rushing through an airport to board a
plane for Santa Barbara and a workshop on nature, religion, and science. As I jogged past a bookshop, a display of *Time* magazines stopped

me in my tracks: the cover pictured a woman, eyes closed and hands clasped in prayer, with a double helix superimposed on her brain. The helix morphed into an image of praying hands.[2] "The God Gene," the cover proclaimed: "Does our DNA compel us to seek a higher power? Believe it or not, some scientists say yes."

Both *The God Gene* and *Religion Explained* exemplify the trend to reduce an evolutionary perspective to one heavily genetic. The material in these books is in keeping with a trend in contemporary science, in which genes are trotted out to explain everything from why your son is shy at school, and why your daughter is a thrill-seeker in sports, to why any young child comes out with long and complicated sentences that no one has ever taught her how to say. Genes are said to explain why people in America's cities and in Africa's Congo Basin kill each other, and why members of different races may tend to be suspicious of each other. Genes explain why men and women tend to act differently when they participate in mating rituals or fall in love. Gene obsession, I call it—genes are sexy, genes are little packages of powerful DNA, and in our culture, also little packages of powerful, media-ready sound bites.

Gene obsession of this sort, like any obsession, results in tunnel vision.[3] Tunnel vision is precisely the outcome when theories insist that humans developed a religious imagination because their DNA is coded in a certain way, or because their brains are triggered by ancient selection pressures in ways that cause ideas of the sacred to emerge. Belongingness, if acknowledged at all in theories like these, is just background noise, a byproduct of the fact that humans live in social groups. What's missing is what matters most: the art and beauty of mutuality, the *force* of belongingness. When people come together in a web of belongingness, they imagine their past and create their future. Drawn to the mysteries of the universe, they relate emotionally to supernatural beings wrapped up with those mysteries, even as they relate to each other.

Why, then, devote a whole chapter to genetic theories? Sometimes, what is missing speaks to us as loudly as what is present. By leaving belongingness out of the story, genetic theories throw into brilliant relief its significance for understanding the origins of religion.

SPIRITUAL GENES

The God Gene is a good place to start. Despite the book's brash title, some major-league backpedaling has appeared by page 8: "There are probably many different genes involved, rather than just one," Hamer admits. "And environmental influences are just as important as genetics." What are we to make of this immediate retreat? Does Hamer mean to embrace belongingness, after all?

The science writer Carl Zimmer, for one, is not satisfied with Hamer's early disclaimer. He suggests a better, more accurate title: *A Gene That Accounts for Less Than One Percent of the Variance Found in Scores on Psychological Questionnaires Designed to Measure a Factor Called Self-Transcendence, Which Can Signify Everything from Belonging to the Green Party to Believing in ESP, According to One Unpublished, Unreplicated Study.*[4] Zimmer's playful revision invites us to unpack the content of *The God Gene.*

The book's argument goes like this. Spirituality is an instinct, and all about self-transcendence, the capacity for experiencing the self at one with the world. Spirituality can be measured by solid detective work in the form of a two-part scientific process. First, people fill out questionnaires designed to collect evidence about their self-transcendence. Then, they submit to genetic testing that reveals to scientists whether the expression of self-transcendence is inherited.

Religion, by contrast, is not instinctual, and cannot be measured in the same way. Religion (or religiousness) amounts to belief in the supernatural, and is expressed through church-related activities connected to that belief. "If our intent had been to measure religiousness rather than spirituality," Hamer writes, ". . . [w]e might have explained how often people attended religious services, for example, or whether they took their children to Sunday school."[5]

Are these definitions on target? Recall, from Chapter 1, that for most people alive in the world today, religious sensibility is boundless and seamless. It *is* life, not a slice of life that can be separated out and measured by a series of objective questions and tests. Prehistoric peoples lived their religion just as much as people in nonindustrial societies do today. Theorizing about religion is ill-served when, from the very first, the subject is boxed off and isolated from the flow of daily life.

Does spirituality fare any better? Is "oneness with the world"—self-transcendence—a reasonable stand-in for spirituality? The reasoning is this: numerical data support the claim that self-transcendence is a "valid psychological trait," one comprising self-forgetfulness, transpersonal identification, and mysticism. Each of these three components has, in turn, its own definition and set of associated questions. The big payoff here, once again, is a numerical score—or, more precisely, one score for each of the three components. Factor analysis, a statistical technique, reveals that the questions for each subscale cohere to measure something real—that is, a good correlation exists among the three subscales in getting at self-transcendence.

Again, let's take stock. Even if we buy the argument that spirituality and self-transcendence are equivalent, other questions loom. Is it valid to collect data on self-transcendence by asking people to reflect upon their own preferences, inclinations, and behaviors? Can people, even people who wish to be honest and disclose all sorts of things about themselves (negative and positive) to strangers, reliably assess their own behavior? When asked to report something about how they are *perceived in the world*, can they intuit how others see them?

So far, there's good reason to feel uneasy with *The God Gene*. The complexities of religion and spirituality fade away when each is defined so simply, and are lost entirely when spirituality is measured by cut-and-dried questionnaires and statistics. And this leaves the gene-testing part still to consider.

Previous studies of twins purported to show that spirituality—but not religion—is significantly heritable, or passed on through the genes. Self-transcendence scores of identical twins (sharing 100 percent of their genes) exceeded those of fraternal twins (sharing, as all siblings do, 50 percent of their genes). With this knowledge as background, Hamer tested 447 pairs of siblings on the self-transcendence scale. He found a higher-than-expected correlation within pairs, meaning that siblings are more similar in self-assessed measures of self-transcendence than they are expected to be.

Here Hamer runs into serious trouble. He is forced to concede "that this test cannot *prove* heritability, since a positive correlation could, in principle, be due to genes, shared environments, or a combination of both."[6] Well, yes. This flaw is more than pesky; it is fatal. Why this is so becomes more clear as we dig deeper into *The God Gene*.

Obtaining DNA samples from some of the siblings who took the self-transcendence test, Hamer isolated differences and similarities in the genetic makeup of his subjects. Previous scientific study came in handy once again when research on the VMAT2 gene came to Hamer's attention. The VMAT2 gene manufactures a protein that packages chemicals called monoamines. Monoamines, including serotonin and dopamine, help regulate the emotions that we feel every day. By packaging the monoamines in a specific way, the VMAT2 gene prepares them to be acted upon by the brain. The take-home message to retain about the VMAT2 is that it plays a role in the brain's emotional system.

Everyone has one or the other version of the VMAT2 gene: one or the other of two "alleles," or alternative forms, of the gene. Hamer's subjects were sorted into two groups, based on which VMAT2 allele they possessed. The next step was to discover whether one of the two alleles correlated better than the other with high scores on the self-transcendence scale. "We hit pay dirt," Hamer says. One of the two alleles—the "spiritual allele"—correlates with self-transcendence, so that people carrying that allele are likely to score high in that trait.

But we're still not done. In an attempt to tease out the role of genetics in spirituality, and to control for influence of the environment, a sib-to-sib comparison was carried out. Siblings have the same economic and racial-ethnic backgrounds, Hamer points out, and they also "go to the same schools and the same churches. They even have the same parental and grandparental influences." Hamer assumes that for siblings, the environment is pretty much constant. This assumption is wrong, and we will return to it, in a big way, soon.

Siblings of the same sex were the ones compared. Hamer thought that most brothers (or most sisters) with the spiritual allele would score higher than their brothers (or sisters) who had the alternative version of the gene. Of the 106 pairs of same-sex siblings who had different VMAT2 alleles, 55 fit the hypothesized pattern. That means in 55 pairs, the sib with the spiritual allele scored higher on measures of spirituality—evidence, according to Hamer, that genes matter. The spiritual allele may alter how the brain goes about processing aspects of experience related to spirituality, and how those experiences actually feel to a person.

Live by the numbers, die by the numbers. It's obvious right off that the numbers indicate a weak trend, and to his credit, Hamer admits this: that 55 of 106 subjects displayed the predicted correlation (between the spiritual allele and high measures of spirituality) translates to just a tad over half of the sample pool. What comes next, though, amounts to the dropping of a bombshell: as it turns out, the spiritual allele raised self-transcendence scores by less than 1 percent of the total variance in the sample population. Hence, Carl Zimmer's mock title.

THE ROLE OF GENES

Genes occupy the center ring in the spiritual-allele scenario. The power of the environment, more specifically the power of belongingness, is neglected. Yet precisely because of belongingness, siblings cannot be an effective control for the role of environmental influence on spirituality. How many of us know families in which two same-sex siblings developed along different emotional paths because they experienced different life events within their family, in school, or in the community?

Consider two sisters, born five years apart. Much can happen within a family over the course of five years. Perhaps a beloved grandparent dies, leaving the younger child without the same treasured spiritual influence that the older one enjoyed. Parents may separate, divorce, remarry, or simply shift in their ways of relating, so that the emotional tenor of the entire household changes. A working parent may lose his or her job, requiring one girl to work many more hours than the other as a teenager. Events such as these may have cascading effects, altering the time or inclination a girl has available for church, meditation, or nature-related experiences.

Or perhaps no such dramatic change ensues. Perhaps, simply, one of the girls comes to admire and trust an inspirational teacher or friend who is deeply interested in spiritual matters—or one who is completely indifferent to them. Any of these factors may influence a girl's spiritual development in profound ways.

Remember that according to *The God Gene*, "spirituality" is genetic, whereas "religion" is based on culture and traditions. But this idea, too, collapses under scrutiny. For one thing, the old dichotomy between nature and nurture is deservedly dead and buried. Asking "How much is nature and how much is nurture?" about any aspect of human behavior (or indeed of animal behavior) is an empty exercise. Genes and the environment are deeply entwined, with enormous influence going both ways. (Gilbert Gottlieb makes this case beautifully in his book *Individual Development and Evolution*.) Hamer cites a study in which it was found that the orthodox beliefs and practices of twins in Indiana were influenced 61 percent by cultural learning and 10 percent by genes. Instantly, the study can be recognized as a case of outmoded science; its conclusion simply fails to make sense.

Hamer concedes that spirituality is not hardwired, because "practice" can strengthen one's spiritual sense. But spicing a genetic theory with a dash of learning (in the guise of practice) fails to save it. In *The*

God Gene, spirituality is defined, after all, as connectedness with something beyond the self. How could a connection beyond the self fail to be affected by belongingness? How could a strong social-emotional connection with another person, no matter one's genetic make-up, not make a difference?

Let's return to our hypothetical sisters. Is it reasonable to expect that experiences with relatives, friends, and mentors would influence only the specific religious beliefs and churchgoing practices the girls might adopt? Isn't there every reason to think that these very same loved ones could influence the girls' spirituality far more than the weak VMAT2 gene could?

Jane Goodall, the world's foremost expert on chimpanzee behavior, lectures and writes frequently about humans' connection to other animals. She conveys the desperate need for people to work to save chimpanzees and other apes from extinction. But for Goodall, a true ethical stance toward animals encompasses small acts in our everyday lives, and not just heroic efforts to save endangered species in Africa and Asia. To refrain from needlessly squashing a spider in the bathtub, to interrupt a busy life long enough to transport an injured bird or turtle to a vet or wildlife rehabilitator . . . these acts make a difference in our world.

In the way she has stressed the continuum of all life on our planet, Goodall has changed my outlook. Through her passion, I have more keenly felt the link between humans and other animals. By taking baby steps such as adopting cats and rabbits that might otherwise face euthanasia; by helping to manage a colony of wonderful and healthy, but homeless, cats; by escorting flies and spiders out of my house rather than squashing them; by moving turtles out of the middle of the road to safety at the side; and by ensuring that injured birds and small animals find their way to rehabilitators, I try to honor that link.

Could this shift in emotion and practice have come about because of my genes? Do I have a spiritual allele that prepared my brain to process Goodall's words in a certain way? I have not been genetically tested. It is possible that I do carry the "right" allele, the spirituality allele. Certainly I carry genes that influence some of my behavior and some of my choices in life: it would be folly to construct an evolutionary perspective that ignores the role of genes altogether. Any behavior that has evolved, by very definition, has a basis in genetics. When we inherit half of our genes from each of our biological parents, we inherit blue eyes or kinky hair. But we do not inherit a gene "for" shyness or a gene "for" depression or a gene "for" spirituality; at most, we may inherit a tendency to express shy qualities or aspects of depression or spiritual yearnings *when the environmental context—including the expression of belongingness in our lives—is ripe for that expression.*

A genetic explanation for the kind of spiritual shift I am talking about vis-à-vis Goodall's influence is deeply inadequate. Why does Goodall's passion inspire thousands of people across the globe to join in grassroots conservation? Is it because these people, too, have the spiritual allele? Could it be, instead, that we humans, so tuned in to belongingness, might be transformed by another person's passion, no matter which version of the VMAT2 gene we have?

Belongingness is downplayed in *The God Gene*, while genetic influences are inflated. Surely science can do better. In fact, serious scientific backlash to gene obsession has begun in full force. The gene is being dethroned by heavyweight scientists such as Evelyn Fox Keller, who points out that the gene is part of an ever changing system in the cell, as much constrained by its environment as capable of influencing its environment. The very existence of a gene "is often both transitory and contingent, depending critically on the functional dynamics of the entire organism."[7]

A cadre of fellow scientists has ardently risen to the defense of the gene. One critic calls Keller's book *The Century of the Gene* "a jihad against our notion of the gene," strong stuff that conveys the level of emotion (admittedly, on both sides) in gene-centered discourse.[8]

Keller is right, though. Occasionally, a single gene may so alter the workings of the body (as in sickle-cell anemia, Tay-Sachs disease, or porphyria) that it is correct to think in terms of a gene-caused disease. By contrast, in the arena of human behavior, little will be accomplished through a search for underlying genetics. It's not as if genes set the blueprint and the environment merely tweaks the blueprint.

Sometimes, gene-enamored thinkers respond to a critique such as mine by trotting out an oft-feared charge, that of "confusing the levels of analysis." Why can't I see that at one level of understanding applied to human behavior, genes rule, and on another, the social environment rules? Can't all this bickering be resolved by simply embracing—and willfully switching between—these two levels of analysis? The answer is no. As I have explained through presenting my own ideas about primates, Greenspan and Shanker's ideas about emotional signaling, and Dissanayake's ideas about human mutuality, people co-construct their experience of the world. One partner may take a leading role, as when a parent guides a child. Co-construction does not demand symmetry, and even when symmetry is absent, each partner may produce change in the other.

So it is with genes and the environment. Growing appreciation exists for the idea that behavioral change can be primary, and can lead to genetic change. Naturally, I do not intend a revival of the theory of the inheritance of acquired characteristics. A person who survives a threatening illness to become unusually empathetic toward others who are sick will not produce a child who has inherited that specific behavioral tendency, any more than a giraffe that has stretched its neck to feed from the branches of a tall tree will produce offspring with longer necks: Lamarckism is discredited for good reasons rooted in the mechanisms of gene transmission. In order to see clearly what I do mean by asserting that behavioral change may precede genetic change, let's revisit the ideas of Greenspan and Shanker from a slightly different perspective. These authors say that certain types of interactive nurturing in a child's life lead to an ability to separate intense emotion from immediate action. The crux of their idea is that through development, and indeed through evolution, there can be a feedback loop between nurturing patterns and symbolic thinking. "As emotional signaling becomes more complex and regulated, it leads to further and further separation of perceptions or images and their originally, relatively fixed emotions and actions. *In this way, complex regulated emotional signaling leads to more and more differentiation between the symbol and its external points of reference, allowing for greater and greater interiorized operations.*"[9]

What Greenspan and Shanker are saying is that in development, and indeed in evolution, change in the quality of emotional communication between social partners can lead to changes in patterns of behaving and thinking. Are genes absent from this picture? Of course not. It takes no stretch of the imagination to see a role for the genes, because enhanced emotional communication, social cohesion, and ability to symbolize all may affect the processes involved in mating and reproducing. The important point is that the change happens because of dynamic shifts at the holistic level of the two-person dyad or the social group.

RELIGIOUS BRAINS

If science can do better than gene obsession, *does* it do so in *Religion Explained?* Spiritual alleles are nowhere in evidence here, and religion is understood to be complex and variable in its expression. Naturally, given my love of anthropology, I'd like to say that it's the author's anthropological training that makes a difference here. It was to Boyer that I turned in Chapter 1, when offering lessons about the diversity of religious expression around the world. Religion *is* emotional in *Religion Explained,* and people do feel intensely when they relate to God, gods, or spirits. Recall the Fang people of Africa, whose emotions were so stirred by the actions of ghosts who gathered at banquets and plotted their next moves. An appreciation for real people living their religion every day may be derived from Boyer's writings, which convey something of the emotional relationship people may experience with the supernatural.

But there are many different brands of anthropology, and the anthropology in *Religion Explained* gives short shrift to belongingness. For Boyer, emotions are not created when people come together in love or hate, in everyday profane practices and everyday sacred practices. Emotions are created and experienced within each person's head. Feelings are "the outcome of complex calculations that specialized systems in our minds carry out in precise terms."[10] And what this means is that religion itself is a "mere consequence or side effect" of the functioning of the human brain.[11]

At first, a categorization of religion as a side effect seems quite startling. Can the abiding love people have for God, or even the deep anxiety people may have about ghosts and spirits, be reducible to a side effect? But a side-effect claim is not necessarily dismissive in and of itself; everything depends on *why* religion is considered a side effect.

Boyer's explanation for religion is closely allied with domain-specific brain models. Evolutionary in nature, to some extent these brain models will sound familiar: Selection pressures in our early prehistory sculpted key brain features. Following the split between the African ape and hominid lineages, human ancestors lived under environmental conditions and in social groups that were relatively stable in terms of their major features. Over millions of years living under these stable environmental and social conditions, our brains adapted specifically, to precise conditions.

But domain specificity goes further than your garden-variety evolutionary perspective in a couple of ways. First, domain specificity is about fixed output from our brains, based on ancient input. Even our twenty-first-century brains are still constrained by these evolutionary forces. Second, prehistoric selection pressures worked in a highly specific way to produce a brain composed of a series of modules, each adapted to solve a particular problem. It's not just that the need to avoid hungry predators selected, in a general way, for human visual acuity (to spot predators at a distance or when they are partly concealed), fast reaction times (to run away), and strong discriminative abilities (to distinguish a dangerous carnivore from an innocuous herbivore). It's not just that the need to get along in social groups selected for cooperation (for people to join together and hunt large game or build big structures) or for an investment in group identity (for a strengthening of "us-versus-them" behavior in aid of survival). No, the pressures were numerous and specific. As a result, the mind evolved to resemble a Swiss Army Knife, with the modules as distinct in what each is meant to accomplish as are the blades of the knife.

Evolutionary psychologists enjoy a guessing game, based on their knowledge of prehistory, about which "blades" should exist in the modern human brain. One such prediction, famous in academic circles, includes no fewer than sixteen modules. This list includes one module each for face recognition, spatial relations, rigid object mechanics, tool use, fear, social exchange, emotion perception, kin-oriented motivation, effort allocation and recalibration, child care, social inference, friendship, semantic inference, grammar acquisition, communication pragmatics, and theory of mind. It even ends with an open-ended "and so on."[12] These modules are seen as relatively impervious to outside influences, a point that is fundamental from a belongingness point of view, and fundamentally wrong.

Let's say your reading of this chapter is interrupted by a ringing telephone. You pick up the receiver and hear the voice of a friend—not an extremely close friend, yet someone who is more than a mere acquaintance. She asks you for a ride to work in the morning. You hesitate; this request is unwelcome. The friend works in a location well out of your way, and ferrying her lengthens your commute significantly. This is the fourth week in a row she has requested the same favor, without the slightest hint that reciprocation may be forthcoming. You wouldn't mind an offer of a return ride one morning, or even a small gift to acknowledge your expenditure of time, effort, and gas money.

Yet breaking through these concerns is awareness of a certain stressed quality in your friend's voice. She has faced some troubling family issues lately, related to caring for a teenage daughter and an elderly mother. Because you empathize with this situation, and recognize that her friendship is worth a bit of effort, you agree to her request. The call ends with warm feelings on both sides.

In the span of a two-minute telephone conversation, you activated and relied upon at least four blades of your brain's Swiss Army knife: your friendship module (you might well have responded differently had repeated requests for favors come from a mere acquaintance), your social exchange module (despite the friendship, a keen awareness of her failure to reciprocate your repeated favors remained intact), your emotion perception module (you intuited an emotional state underlying her words), and your theory-of-mind module (you were able to put yourself in your friend's shoes and empathize with her family troubles). Almost certainly, even more modules than these four were used, because you relied on skills of communication (ranging from semantics to grammar) as you conversed with your friend.

But look at what goes missing in this account: the very same social and emotional influences so strikingly absent from *The God Gene*. All the brain modules in the world will not sway a person who has been raised to value tit-for-tat economics over loyalty in friendships, or a person whose family taught her to share resources only with blood relatives and not "outsiders." Of course, such a person may grow up to reject the principles with which she was raised. If she does, it would probably be because she learned about a different way to be, from a friend, teacher, mentor, or spiritual figure.

Mental-module advocates admit that variation in human behavior exists, but say that it is all filtered through a hardwired brain. In *Religion Explained*, inference systems of the brain hold the key to human behavior. Inference systems are specialized processors that deal with intuitive psychology, moral issues, goal detection, and so on. Like mental modules, inference systems cause our modern brain to solve ancient problems: "Evolution," says Boyer, "does not create specific behaviors; it creates mental organization that makes people behave in certain ways."[13]

In thinking about our evolutionary past, Boyer focuses on hominids as both predator and prey. Hominids must get enough to eat, but must also avoid being consumed by the larger meat-eaters all around them. Further, each individual hominid lives in a small group, surrounded by other hominids who are possible rivals or allies in the endless quest for resources. In this situation, knowing who around you "knows what, who is not aware of what, who did what with whom, when and what for" can confer an edge in the evolutionary game of survival and reproduction. Because of these strong selection pressures, humans evolved with a hair-trigger detection system for *agency*—that is, for some "agent" alive in the world whose actions can affect us. With agency we arrive at the crux of Boyer's model, where he brings together the concepts of "overdetection" and "decoupling."

For vulnerable hominids, an agency-detection system that is easily activated conveys multiple benefits: when it comes to agency, far better to overdetect than to underdetect. It's preferable to see possible harmful agents where none exist than to miss predators all too ready to consume you or your family members. "The expense of false positives is minimal," as Boyer says. This bias toward overdetection fits with a second capacity, one that allowed hominids to make inferences about agents not directly present. Being able to infer what an as-yet-unseen predator might do could be highly beneficial, as could figuring out the possible strategy of a rival who is temporarily absent from the group. So the human mind evolved to make inferences decoupled from the immediate here-and-now.

Now we are in a better position to grasp the explanation offered by *Religion Explained*. The human religious imagination arose because the hominid agency-detection system was highly prepared to interact with perceived agents-at-a-distance. Rejecting the notion that humans invented gods and spirits, Boyer insists that people "receive information that leads them to build such concepts."[14] Supernatural beings, after all, are ideal exemplars of "decoupled agency." And people everywhere think of God, gods, and spirits as active in their world, as agents of change, which affect their own lives directly. That some religious ideas commonly occur (the souls of dead people can visit the living) whereas others never occur (God is omniscient but powerless) can now be understood. Our religious ideas are constrained because they are byproducts of our specialized inference systems: "We have the cultural concepts we do because the way our brains are put together makes it very difficult for us not to build them," as Boyer puts it.

The upshot of all this is that God, gods, and spirits exist for us *only* because our brains detect them, which occurs in turn *only* because of the features of our ancient mental architecture. Boyer explains: "The intuitive psychology system treats ancestors (or God) as intentional agents, the exchange system treats them as exchange partners, the moral system treats them as potential witnesses to moral action."[15] Inference systems then explain not only where religious beliefs, practices, and emotions come from but also their meaning.

Just as *The God Gene* downplays the dynamic relationship between genes and the environment (including belongingness), so *Religion Explained* downplays the dynamic interaction of the brain with the environment (including belongingness). Just as it isn't convincing to assert that *The God Gene* really does embrace the role of the environment, so, too, is it with *Religion Explained*. This point is important enough to revisit: theories of inference systems, to which brain architecture is so critical, do not omit consideration of the environment, but they fail to take into account the *creative force* of belongingness within the environment.

Current theorizing about God and the origins of religion is highly prone to domain specificity.[16] Far more convincing than domain-specific models is evidence from neuroscience about the amazing malleability of the human brain.

PLASTIC BRAINS

Experience rewires our brains, as it does, to some degree, the brain of any mammal. Scientists were jolted to learn the extent of brain plasticity when, in the 1940s, the psychologist Donald Hebb filled his house with rats. Placing these rodents in a complex environment far more stimulating than the typical laboratory setting, Hebb demonstrated that enriched experience changes the ways rat brains grow and function.

What do we know about plasticity in the human brain? At one level is synaptic plasticity. Synapses are the places in the brain where neurons (individual cells) communicate across small gaps, by way of the very same neurotransmitters involved with Dean Hamer's VMAT2 gene.

Communication between particular neurons can be strengthened by experience; learning thus changes the brain. It also appears now that entire networks of neurons—indeed, whole brain regions—may be similarly affected by experience.[17]

Consider the case of Superman. The American actor Christopher Reeve became a virtual poster-person for brain plasticity following an accident during an equestrian competition in Virginia. Falling from his horse, Reeve sustained a serious spinal cord injury that left him completely paralyzed from the shoulders down. The prognosis offered was permanent paralysis. Indeed, at first Reeve was unable to move his limbs even a little, and he could not breathe unassisted. Nonetheless, he began to exercise tirelessly, with the goal of walking again.

Stunningly, Reeve regained limited movement in certain parts of his body. After discovering that he could move an index finger entirely on his own, he went on to recover motion in each of his ten fingers. Floating in water, Reeve was able to move each joint. Lying down, he could push hard with each leg against his therapist's chest.[18]

Through frequent exercise, Reeve reanimated the dormant connections between his spinal cord and his brain. Both Reeve's persistence and his financial resources were exceptional, but the results he achieved are not unique. Hundreds of people, lacking full sensation in their lower bodies if not as severely paralyzed as Reeve, are now able to walk after trying out techniques similar to his.[19]

In some cases, the human brain actually reorganizes itself. When a blind person reads Braille, it is the *visual* areas of her brain that respond. As she touches the bumps on the page, her visual cortex operates just as yours is doing now as you read these words. Recruited for the task at hand, the brain's visual area devotes itself to a new task.

Though brain plasticity is often described in terms of what follows after a trauma—an accident or a stroke, or a person going blind— nothing so catastrophic need occur. The brains of musicians are highly adaptable. One study found that the auditory cortex in musicians is 130 percent larger than in nonmusicians; the degree of increase was correlated with the extent of musical training. Another study, this one of violinists, revealed that the regions of the brain receiving input from the left hand (but not from the right) were significantly larger in musicians than in nonmusicians. It's the left hand that makes complex movements when the violin is played, so this finding amounts to a discovery that relates brain changes directly to experience.[20] Here, then, is scientific backing for all those clichés beloved by parents who urge their children to practice, practice, and practice their instruments: practice not only makes perfect, it also makes a rewired brain!

Brain studies show that different areas of the brain cooperate across what are supposed to be separate modules. The crashing tide of evidence argues against the picture, popular though it is, of a modern-day brain either wired to our past, or carved up into separately functioning modules. Further, in claiming that humans have complex social relationships because we have the specialized mental capacities that social life requires, *Religion Explained* has it exactly the wrong way around.[21] Belongingness shapes and structures our very existence, and indeed, is the "driver" of the cognitive capacities that underwrite our complex emotional interactions, as is explained by Greenspan and Shanker's *The First Idea*.[22]

Bucking the trend of gene obsession and brain domain-specificity, *The First Idea* shows how certain types of caregiving practices account for certain universal human behaviors. When infants and caregivers communicate together in deep emotional connection, the developing child is pushed toward greater and greater emotional control and calm, reflective thinking. Change over evolutionary time, too, can be explained by ever more sustained bouts of emotional exchanges. And as we know, belongingness has its roots well back in the primate lineage.

Religion Explained, then, takes us a little farther than *The God Gene*—but not much farther. Each exemplifies a watered-down approach to religion, as I have discussed at length. It is worth pointing out, too, that both diverge (though in different ways) from an adaptationist approach to religion.

Adaptationists want to know how hominids who embraced religious ideas or practices might have fared better than hominids who did not. If abilities like swift running and acuity of vision might have helped some hominids outcompete others in the game of survival and reproduction, could the human religious imagination have played a similar role?

Fear of the unknown is the single most suggested reason for why religion developed in the first place. Hominids lived in a world of mystery. Unpredictable cataclysms ranging from storms to drought to earthquakes must have been random and incomprehensible events for australopithecine mothers carrying their babies across the African savanna, for *Homo erectus* groups trekking out of Africa, for Neandertal men and women burying their dead companions. Death from sudden illness or the complications of childbirth, together with other smaller-scale and apparently capricious tragedies, must have loomed as fearful possibilities. In response to a massive species-wide anxiety attack, then, came the idea that some powerful being or beings were in control, and could be placated if certain actions were performed with care and devotion.[23] So were born ritual and religion.

Generations of evolutionary scenarios have embraced this adaptationist idea or something like it. Hominid individuals, or perhaps entire social groups, that "got religion" were better able to compete successfully. Members of these groups came together around a belief system, a coherent framework from which to deal with life's mysteries and stresses. If not always able to provide escape or relief from those stresses, such belief systems did provide collective comfort from various terrors, including fear of death.

In *Religion Explained*, Boyer insists that "a religious world is often every bit as terrifying as a world without supernatural presence, and many religions create not so much reassurance as a thick pall of gloom."[24] Point taken. Still, the story of human evolution is the story of adaptation through group living, and that group living must have been, in part, about dealing with mysteries and worries. Over time, as the archaeological record of Neandertals and *Homo sapiens* attests, hominids invested more and more in group allegiance and identity, whether through particular self-adornment (jewelry, red ocher) or by participation in group rituals. Groups of people can even be understood as organisms, living entities that compete with other such entities to survive and reproduce just as individuals do. Some groups live on to reproduce and others do not.

The group-selection idea is central to David Sloan Wilson's *Darwin's Cathedral*. Firmly embedded in the adaptationist camp, Wilson considers religion to be adaptive in its own right, not a mere hitchhiker freeloading in our brain. His description of social life harks back to Durkheim's writings; people are made to be vital and alive, behaving with emotion (and not emotion generated by a spiritual allele or a well-oiled inference system).

Yet here again, acknowledging that people feel their religion, and express that emotion in groups, isn't enough. When Kahlil Gibran writes, "Your daily life is your temple and your religion," he expresses a fundamental truth gleaned from the anthropology of religion. In people who live traditionally (by hunting-gathering, or by farming and domesticating animals), religion is (and was) about singing, dancing, shamanistic trances, and healing practices. Doing and relating are the bedrock of spirituality, and they *create* spirituality.

Too many modern evolutionary accounts of religion have lost a nuanced sense of what it means to be wholly social beings.[25] In many theories, a "sexy" reductionism is visible, that is, a relegating of the social dimension to a mere variable, coupled with an inflation of the power of genes or a view of the brain as composed of specialized and fixed "blades."

RECOGNIZING WRONGNESS

Scientists are honor bound to try to figure out whether their ideas are wrong. A strong temptation exists to seek evidence only in support of one's ideas, and not evidence in opposition to them. Rarely does a scientist set out to twist the evidence or to deceive anyone outright. Instead, the temptation may be part of a larger human tendency to find validation for one's assumptions, even in the face of evidence to the contrary.[26]

Budding scientists are taught from the very moment they don their metaphorical white coats that the scientific process is keyed to *falsification:* that is, predictions based on the scientist's ideas must be tested by observation and/or experimentation in the real world, and found false whenever possible. As counterintuitive as it may sound, the ideal is for a scientist to find herself wrong.

Testing ideas and supporting them with evidence is always the gold standard in science, a principle that holds as true for the study of prehistory as for anything else. But here we have to be more resourceful, and not be defeated too easily when testing and evidence-gathering run into limits. The ideal of prediction-testing should not be weaponized by skeptics. Any new set of ideas about our past that involves untestable speculation should not be dismissed automatically: there's no reason to require that the standard of immediate falsifiability be the ultimate one for all reconstructions of prehistory.

To assert that prediction-testing is an enormously difficult enterprise when dealing with something intangible, such as the religious imagination and its evolution, is merely to rehearse material covered in earlier chapters. It's challenging enough to trace the prehistory of things written in bone (bipedalism, for instance) or stone (development of the first nonorganic tool technology), but when we are dealing with meaning and emotion about the sacred in our past, our ability to test predictions against the real world is seriously challenged.

The "real world" of the prehistory of religion is best recoverable by us through material culture, including tools and other artifacts; through art; and through the use of primate studies, anthropology, and psychology that are grounded in evolutionary theory. Collectively, these sources help us understand the evolution of belongingness. For earlier stages of prehistory, analogy with living apes is most useful. For later stages, understanding may come from ethnographic analogy as well, that is, from analogy with practices in living societies in the world today. Reconstructions that carefully integrate material from these various sources have a high probability of leading to explanations that shed light on the origins of religion.

And here we come to the bottom line: Hominids turned to the sacred realm because they evolved to relate in deeply emotional ways with their social partners, because the resulting mutuality engendered its own creativity and generated increasingly nuanced expressions of belongingness over time, and because the human brain evolved to allow an extension of this belongingness beyond the here and now. All of these things were necessary for the origins of the human religious imagination.

Points made by the linguist Tom Givon in an online discussion are helpful here. Though Givon writes about the scientific process as it relates to the origins of language, his comments mesh perfectly with what I am saying about the origins of religion:

> When we discuss language evolution, we are honor-bound to concede that making testable predictions is—at the moment—hard to come by, much harder than in physics or biology. . . . I think that some day we will be able to make testable predictions. But they are not going to be exactly of the same kind as predictions one can make about "hard-wired" bio-evolution. . . . At the moment we must concede we are at the hypothesis-formation stage, groping for a theory that WILL make predictions. . . . And since when is difficulty a reason to quit?[27]

What I like about Givon's comment is its unapologetic combination of realism and optimism. At the moment, we cannot test all the predictions we might like to test about the expression of hominid emotion, or the meanings that hominids may have created as they acted ritually with symbols. We cannot test very many predictions at all about the development of spirituality in hominids' everyday lives. But this is no reason to back entirely away from the goal of offering explanation about the origins of religion.

Some of the ideas presented in this book are testable at present. We can look for more Neandertal burials to be uncovered that have symbolic markers associated with them. Caves used by early *Homo sapiens* should include built structures like the one found at Bruniquel in France and dark, inaccessible areas of caves should show signs of hominid use. When new hypotheses are constructed about the origins of religion, they must be sensitive to the rich evolutionary platform for hominid emotional engagement with others. Our hypotheses are only as good as the notions on which they are based; genetics and fixed, modular brain architecture leave us stuck at the starting line. The story of apes' and hominids' lives is too stunningly emotional, and creative, and full of mutuality, for us to remain satisfied with that.

In the future, scientists will certainly learn more from direct archaeological research about hominid belongingness, symbolic ritual, and spirituality. Testability and falsifiability will continue to be the primary goals. But for now, it may be enough that a set of ideas be highly compatible with what we know of the archaeology of human evolution and the anthropology of traditional societies; with evolutionary theory; and with the evolutionary platform from primate studies—or, more correctly, *more highly compatible* with these than are competing theories.

EIGHT

God and Science in Twenty-first-Century America

R EADING AND WRITING intensively about the human religious imagination, I became deeply curious about its expression right on my home turf, semi-rural Gloucester County, Virginia. Gloucester is located slightly north of the famous Yorktown battlefield, where in 1781, General Lord Cornwallis surrendered to General George Washington and brought the American Revolution to an end. Running through this area today is a thoroughfare called Route 17.

One day, I drove the mile or so from my home, on a private road, out to this highway, then continued three miles north, made a U-turn, and came back the same distance on the other side of 17. I photographed every church I saw along the way.[1] No fewer than twelve, in fact, can be found in this stretch of road, ranging from tiny storefront locations for religious gatherings to a magnificent 350-year old Episcopal church, the largest colonial church in the state.

Each church displayed its name prominently. Some also posted short Bible verses, or the times at which services are held. Others advertised themselves by way of a portable roadside sign, the kind a fast-food restaurant might use. In place of "Double Burgers, Biggie Fries 2 / 3$!," a church posts a mini-catechism pithy enough to be read by hurried motorists. On the day I surveyed, a sign at the Church of the Way and Truth asked, "God is having a family reunion are you going to be there" (readers are left to supply punctuation). A sign at the New Life Church of God proclaimed, "God's love is a place you can always call home." And the First Baptist Church announced, "Free trip to heaven. Details inside!"

This set of a dozen churches is not diverse. No Jewish, Muslim, Mormon, Buddhist, or even Catholic place of worship is represented. I live in the American South, after all, and I sampled in a tiny area, well away from a city. In fact, I was surprised at the sheer number of options available for people who want to worship as part of a congregation.

We are a country saturated with all things religious, at least when "religious" is taken to imply preoccupation with God. Indeed, I want to retreat from a broad, anthropological view of religion in order to consider life in the United States solely according to matters of faith and belief in God. How can we understand the religious climate, or more pointedly, the perceived relationship between religion and science, in this country at this particular point in its history? How prepared are Americans to think about the origins of religion in an evolutionary perspective, as this book asks its readers to do?

AMERICAN VIGNETTES

Perhaps it is an irony that, in our high-speed, computer-age nation, religion is woven through so much of everyday American life, just as the sacred was part of everyday life throughout prehistory. Spilling out from the aisles of churches, temples, and mosques, religion pervades all aspects of American society. Yet this pervasiveness comes with a modern twist, because religion in America is heavily influenced by our fast-paced, bigger-is-better, technologically oriented, and quite politicized culture:

- At Knott's Berry Farm, the classic amusement park south of Los Angeles, visitors exiting one of the numerous thrill rides may proceed straight to Sunday church services. In an area called Ghost Town, park-goers wander among re-created aspects of the 1880s Old West, and then come upon the Church of Reflection. Inside, an evangelical preacher offers two services each Sunday to a small congregation composed of park employees and visitors.[2]

- Fun-seekers who believe that God created the dinosaurs on Day Six of the universe's formation may bypass Knott's Berry Farm and visit Dinosaur Adventure Land in Pensacola, Florida. No thrill rides are found here, but the park's founder, a minister named Kent Hovind, has established a discovery center, museum, and a dozen outdoor games. Each of these in some way teaches visitors about creationist beliefs, including the notion that humans and dinosaurs coexisted.[3]

- In Texas, worshippers may join a congregation that, at 30,000 strong, is a different species of church altogether from the one at Knott's Berry Farm. With a choir of 500 members, and 16,000 congregants who worship together at each service, this arena-based church is the nation's biggest. Thanks to a $95 million renovation in the arena, a pair of waterfalls surrounds the prayerful during the four services offered each weekend, and television audiences join in the worship. Church coffers swelled to $55 million in 2004 alone.[4]

- And thanks to a rabbi based in Hattiesburg, Mississippi, people wishing to convert to Judaism may now do so via the Internet. Rabbi Celso Cukierkorn offers an online curriculum customized for each student. Internet study is capped by required attendance at an in-person conversion seminar. People who live in areas where no rabbi is readily available, or who simply have crowded schedules, can now convert with relative ease.[5]

• Interactive role-playing awaits those who visit the Church of Fools, a cyber-temple sponsored in part by the Methodist Church of Great Britain. Visitors can, through their on-screen personas, "kneel in prayer, talk or whisper in text messages, extend a hand in blessing or raise both arms in ecstatic praise. They can also sit in pews or gather for conversation in a crypt equipped not only with chairs but with a 'holy water' water cooler and vending machines as well."[6] Why include a British church in a section on the American religious sensibility? The majority of visitors to the cyber-church are, in fact, American.

• Even people who haven't read the mega-blockbluster novel *The Da Vinci Code* may know something of its plot, which revolves around a married Jesus, his wife Mary Magdalene, and the Holy Grail. Translated into forty languages, the book has been discussed endlessly in the media, and spawned a cottage industry of spinoff books, documentaries, and academic discussion panels. In a nutshell, the novel, and the movie made from it, have ignited a singular firestorm of debate about the nature of early Christianity.

• God has become a force in presidential politics. While this phenomenon is not new in American history, it took on new dimensions with George W. Bush's presidency. Trying to spot and interpret the religious symbols with which Bush laced his speeches and interviews became a parlor game of its own. One hot question that has been asked in the media: does Bush "double code" his words?[7] Does he use phrases, ordinary-seeming to non-Christians, as tip-offs that he intends to promote fundamentalist Christian beliefs through his policies? When Bush used the phrase "wonder-working power" in his 2003 State of the Union speech, was this just an apt allusion (to a Protestant hymn) or was it more? Some observers even insisted that the image of a cross subtly graced Bush's podium

during the Republication National Convention. Bush ran
against a man who is a devout Catholic—but who kept his
faith a private matter. While it is too simplistic to suggest that
Bush's frequent professions of faith decided the Bush-Kerry
contest, few doubt that it was a factor. Wound-licking
Democrats certainly did not doubt it, for they hired
consultants to help them inject religion into their party
platform.

Of course, the depth and the breadth of ways that faith may be expressed in twenty-first-century America is not reducible to a set of
quirky observations. The deep emotions associated with religion may
be expressed publicly through art, music, and dance, just as they have
been since the days of early *Homo sapiens*. Robert Wuthnow, director of
Princeton University's Center for the Study of Religion, points out
that the arts and religion both "encourage reflection, and both orient
people toward mystery and transcendence." People with greater expo-
sure to the arts, regardless of other factors such as education level,
engage more in prayer, meditation, and other activities that are considered to be spiritual in orientation.[8]

Of course, this raises a chicken-and-egg question; perhaps spiritual
people seek out the arts more. Whatever the cause and whatever the effect, it's clear that creating art may be powerfully spiritual for the person doing the creating. The famous choreographer Martha Graham
referred to dancers as "acrobats of God," saying, "where a dancer stands
ready, that spot is holy ground."[9] Though atheist or agnostic dancers
might object strenuously to Graham's generalization, many dancers do
experience elements of their craft as ritual, and many do feel that they
are "dancing spirit," in the words of one.[10]

Across the land, from prairie towns to inner cities, from seaports
to mountain hamlets, people from a multitude of faiths carry out good
works in God's name, with no fanfare and no limelight. Others resist
the conflation of American culture with American religion; they work
to ensure that freedom of religious expression not override the separa-

tion of church and state. Freedom of religion, they remind us, must include the freedom to reject belief in God or any higher power.

THE RELIGION-SCIENCE WARS

The story of American religion is embedded in a second story, a story driven by intense polarization between religion and science. The political columnist Nicholas D. Kristof refers to this polarization as a *"poisonous* divide." A first clue as to the depth of this divide comes from pollsters who ask a dizzying array of questions related to religion and spirituality. Let's look at some of the numbers:

- 90 percent of American adults believe in God
- 84 percent believe in the survival of the soul after death
- 84 percent believe in miracles
- 82 percent believe in heaven
- 69 percent believe in hell
- 68 percent believe in the devil[11]
- 77 percent believe in angels
- 20 percent believe they have seen an angel or know someone who has seen one[12]

How do Americans respond to questions about an ancient earth and a human lineage that gradually evolved?

- 60 percent of Americans believe that the Bible stories are word-for-word true; this includes the Bible's explanation of Moses parting the Red Sea; God creating the world in six days; and Noah's ark[13]
- 55 percent of Americans believe that God created humans in our present form; 27 percent of Americans believe that humans evolved, but that God guided the process; 13 percent believe that humans evolved without any guidance from God[14]
- 37 percent of Americans want the teaching of creationism to

replace the teaching of evolution in the schools; 65 percent favor schools teaching creationism along with evolution[15]

• In Louisiana, 24 percent of biology teachers believe that creationism has a scientific foundation; a further 17 percent were unsure[16]

In short, six times as many Americans believe in angels as believe that evolution occurred without God's help. This finding is stunning because it emerges in a country that wishes to be on the cutting edge of science. Yet there's every reason to believe the numbers are valid; the figures remain relatively stable year to year, and these specific results match up well with results from other polls. Less than a third of American adults, for example, can identify DNA as the mechanism of heredity, and a fifth say that the sun revolves around the earth.[17] Science is not a strong point of the American populace.

An American who believes that evolution unfolded without God's aid is decidedly in a minority. People who embrace an evolutionary perspective as a way to make sense of the world and their place in it must surely be even rarer! (It's anyone's guess what percentage of Americans embraces the application of an evolutionary perspective to understanding the origins of religion.)

"Poisonous" is, sadly, an apt term for the climate in America today, because religion is continually made to seem opposed to science. Far too often the discussion is starkly, and falsely, set in terms of a dichotomy: "Do you believe in God or do you believe in evolution?" The wording of this inquiry puts faith and scientific knowledge on a parallel footing, but what can it possibly mean to *believe in* evolution? That phrase just doesn't compute; it makes as much sense as asking "Do you believe in gravity?"[18]

This state of affairs cannot be understood without exploring two schools of thought that purport to challenge evolutionary theory, creationism and intelligent design. It is important to understand how these two sets of ideas are alike, how they differ, and why both must be classified as religious rather than as scientific ideas.

CREATIONISM AND INTELLIGENT DESIGN

Challenges to evolution from creationists are as old as the hills—although the hills' age is still disputed. Most creationists believe that God created the earth and all creatures on it at about 10,000 years ago. The billions of years of geological history written into the rocks at the Grand Canyon; the bones of 3-million-year-old Lucy and those of her ancestors and descendants; and all other evidence of evolution's action are explained away. For instance, the biblical story of Noah's ark—which is accepted as literally true by the majority of Americans—is invoked as an explanation for the carving of the Grand Canyon and for the presence of fossils in deep layers of the earth.

Because it is only fair to judge creationism by creationists' own words, I quote some here concerning the effects of Noah's flood:

> Creationist geologists agree that rushing water formed Grand Canyon. Some suggest it was Noah's floodwaters as they flowed off the continent (Genesis 8:3). Others suggest it was a post-Flood regional catastrophe caused when a huge mass of inland water, left over from the Flood (and excessive post-Flood rainfall), suddenly breached its natural restraints and rushed to the sea. The idea that canyons invariably take vast ages to form is unfortunately very firmly cemented in the public mind. Even today, most school students are, regrettably, still taught the older, long-age model of formation for Grand Canyon, for instance.[19]

And: "Only catastrophic conditions can explain most fossils—for example, a global flood and its aftermath of widespread regional catastrophism."[20]

These words reveal a rejection of science and a desire to revamp the material taught in the schools. When creationist views are taught in church, Sunday school classes, or other religious settings, no one need bat an eye. Attendance at such classes is wholly voluntary; the Constitution's First Amendment grants freedom of religious belief to all, and "all" includes creationists. But creationists go further, insisting that their beliefs amount to "creation *science*." They lobby for their ideas to be taught in public schools, by science teachers, right alongside evolution. And here creationists run afoul of the remainder of the First Amendment—the part that the U.S. Supreme Court has interpreted to guarantee, in Thomas Jefferson's phrase, "a wall of separation" between church and state.

Creationism's basic tenets are antithetical to a scientific understanding of an ancient earth and the forces of evolution. Creationists are explicit about their belief in an all-powerful God who created the earth and all its creatures, including humans, in the relatively recent past. This willingness to be explicit is welcome because it offers people a crystal-clear choice: believe in God's creation at 10,000 years ago, or reject that idea; believe that these God-centered views belong in the science classroom, or reject that premise.

The waters become murkier when dealing with a second attempted challenge to evolutionary theory. Unlike creationism, intelligent design (ID) accepts that the earth is ancient, and cloaks religion as science in a different way. ID's central principle is called irreducible complexity: certain natural systems, or organs, are said to be simply too complex to have evolved step-by-step through evolutionary processes. One example is the human eye. With all its intricate and sophisticated working parts functioning together as a coherent whole, how could a half an eye (a partly evolved eye) have been any use to a living creature? It could not, say ID proponents; the eye must have been designed as a complex organ. Some ID advocates name God as the designer, whereas others refuse to specify.

ID proponents try to distance themselves from creationists. Admitting, unlike creationists, that fossils are ancient and that evolution does account for much of what we see in the natural world, they sell themselves as engaged in the scientific process. Yet the claims they make around irreducible complexity fall apart under scrutiny. The eye's evolution is no mystery; scientists have charted its stepwise evolution so cogently as to allow Richard Dawkins to call the whole complex-eye issue "a lightweight question."[21] Primitive forms of the eye, without the type of lens that marks the human eye, were useful to certain species in certain environments; evolutionary forces such as mutation and natural selection acted to modify these forms. The complex human eye is the result of a process of adaptation that has quite ancient roots.

ID's fatal shortcomings are widely documented, and I will not rehearse them in greater depth here.[22] I do want to stress that while the ID cloaking of religion as science is sometimes thick, with no mention of God as the designer, at other times it is quite sheer, with God invoked quite plainly. Here is a comment made by a leading ID advocate, the biochemist Michael J. Behe:

> I'm still not against Darwinian evolution on theological grounds. I'm against it on scientific grounds. I think God could have made life using apparently random mutation and natural selection. But my reading of the scientific evidence is that he did not do it that way, that there was a more active guiding. I think that we are all descended from some single cell in the distant past but that that cell and later parts of life were intentionally produced as the result of intelligent activity. As a Christian, I say that intelligence is very likely to be God.[23]

Indeed, one may wonder, what could be the designing force envisioned by ID, if not God?

An issue central to these "evolution wars" relates to the term "theory." A fundamental confusion exists because creationists and ID advocates use the term one way, and scientists use it quite another way. Here's how I explain the situation to my students each fall as we begin exploring the evolutionary perspective together:

Let's suppose that you return from classes one day to discover your roommate Chris in tears. You try to comfort her, but she does not wish to talk, so you leave to join some friends for dinner. To them you remark, "Chris is so stressed; she won't talk to me so I don't know what is going on. I could see the screen of her laptop, though, so my theory is that Joe broke up with her. She started to cry right after IM'ing [instant-messaging] with him."

In this case, "theory" means an educated guess. Your idea about Joe may be right or it may be wrong. Chris might just as well have been crying over bad news that she had received and shared with Joe, but that didn't involve him directly. Or perhaps Joe found out he had to work over spring break, instead of jetting off to Tampa as the couple had planned. Applying the word "theory" in this situation is quite common, despite the situation's evident ambiguity. Unless you move in science-geek circles, it's unlikely that any of your friends would point out that you had offered a *hypothesis*—not a theory—about Chris's behavior.

In science, a hypothesis is an idea that requires the collection of evidence. Only when it is tested against the real world does a hypothesis carry any credibility. In scientific jargon, a hypothesis must be falsifiable: a way must exist to discover whether it is wrong. The word "theory" refers only to a set of ideas that have been expressed as hypotheses, tested in the real world, and *not falsified, but instead found to have solid support.* That evolution is stamped "a theory" means that it is an evidence-based, tried-and-true framework for understanding the biological world.

By contrast, neither creationism nor intelligent design is a theory. Neither has scientific evidence to support it. Neither is falsifiable. No person can hope ever to test how a God, or some other unspecified intelligent designer, may have acted in the universe. Neither set of ideas poses any credible scientific challenge to evolutionary theory.

The scientific dismissal of creationism and ID does not affect the fact that both—and ID even more than creationism—play an increasingly visible role in American legal and political life. Let's explore ID a bit more deeply.

PRESIDENTS AND JUDGES WEIGH IN

As president, George W. Bush openly stated his view that intelligent design should be taught along with evolution in the public schools. "I think that part of education is to expose people to different schools of thought," Bush said. People, he said, should be exposed to both "a theory of evolution and a theory of creationism," and "I personally believe God created the Earth."[24]

Judging from these remarks, Bush does not distinguish between the colloquial use and the scientific use of the word "theory." Certainly he joins the majority of Americans in accepting a God-centered view of the world. And certainly he's far from the only politician willing to weigh in on the evolution debate. Consider the following excerpt from an op-ed in the newspaper *USA Today*:

> But in this tremendous effort to support Charles Darwin's theory of evolution, in all these "mountains of information," there has not been any scientific fossil evidence linking apes to man. The trouble with the "missing link" is that it is still missing! In fact, the whole fossil chain that could link apes to man is also missing! The theory of evolution, which states that man evolved from some other species, has more holes in it than a crocheted bathtub. I realize that is a dramatic statement, so to be clear, let me restate: There is zero scientific fossil evidence that demonstrates organic evolutionary linkage between primates and man.[25]

The tag line for this op-ed reads: "Utah State Sen. D. Chris Buttars, R–West Jordan, is active on the evolution-education issue." Responding to the op-ed, scientists were incredulous that an elected official, an advisor to educators, could be so out of touch with modern science. Bush's own remarks were widely interpreted as a big boost for ID, as is reflected in headlines from two California papers that reported on Bush's ID-related statements: "Inspiration for Doubters of Darwin" and "Bush Pushes Very Hot Button: President's Comments Embolden Anti-evolutionists."[26]

Blowback to political statements of this sort can be quite acerbic, as when Cynthia Tucker wrote that the United States is "led by a cult of religious fundamentalists who wish to impose their narrow thinking on the rest of us."[27] In voicing its deep disappointment with Bush, the

National Science Teachers Association referred to ID as "pseudo-science."[28]

Accusations of close-mindedness on the part of ID's critics became the blowback to the blowback. The Discovery Institute, a think tank headquartered in Seattle, posts on its website an innocuous-sounding mission statement: "The Institute discovers and promotes ideas in the common sense tradition of representative government, the free market and individual liberty." Further exploration, however, reveals that the institute's Center for Science and Culture (CSC) supports not only the teaching of evolution but also "the work of scholars who challenge various aspects of neo-Darwinian theory and scholars who are working on the scientific theory known as intelligent design."[29] (Note the incorrect usage of the word "theory" here). And the CSC even touts a "free speech on evolution campaign" that urges people to fight back against "malicious" and "fundamentalist" Darwinians: "Across America, the freedom of scientists, teachers, and students to question Darwin is coming under increasing attack by what can only be called Darwinian fundamentalists. These self-appointed defenders of the theory of evolution are waging a malicious campaign to demonize and blacklist anyone who disagrees with them."[30]

The CSC's director, Stephen C. Meyer, admits to ambitious goals: "We are in the very initial stages of a scientific revolution. We want to have an effect on the dominant view of our culture."[31] And the CSC nets cash enough to help foment its preferred brand of revolution. According to the New York Times, the Discovery Institute is "financed by some of the same Christian conservatives who helped Mr. Bush win the White House."[32]

The Discovery Institute's battle cry is "teach the controversy"—that is, teach ID alongside evolution. This issue is now fought, in high-profile ways, in state legislatures and courts, more than eighty years after the famed Scopes "monkey" trial, and twenty years after the U.S. Supreme Court ruled that creationism cannot be taught in public schools along with evolution. The highest-profile case to date was litigated in Dover, Pennsylvania, where U.S. District Judge John E. Jones ruled against the teaching of ID in public-school science class. When he included in his ruling a reference to the "breathtaking inanity" of ID thinking, Jones ignited a firestorm across the nation.

AMERICAN-MADE?

From all that Indian people tell us today, and from all anthropologists can surmise, the first-ever Americans were deeply spiritual people. The populating of America by settlers from overseas, starting at Plymouth, was in part driven by flight from religious persecution. Americans became a nation in a time and place saturated with religious concerns. The American story from the first is a story of belongingness in a religious context. But is this a peculiarly American story?

Poll numbers suggest that the oft-heard phrase "the secularization of Europe" contains some truth. Surveys carried out in fourteen countries indicate that 70 percent of European adults believe in God. This differs appreciably from the 90 percent who reportedly do so in the United States, but it is an average, and by definition it masks variation. In Poland, 97 percent of the population believes in God, whereas in the Czech Republic only 37 percent do. Between these extremes is much divergence: Portugal is at 90 percent, the Netherlands at 51 percent.[33]

A different poll, however, casts a rather different light on the situation. The European Values Study found that religion is "very important" to only 21 percent of Europeans. The discrepancy with the comparable figure in America, 58 percent, is striking.[34] Active church attendance in Europe is declining within every major religious tradition. More than a third of Europeans (36 percent) say that they "never" or "practically never" attend church; less than half as many Americans (16 percent) say this. France leads the list of avowed nonchurchgoers (60 percent), closely followed by Britain (55 percent). Almost identical to the American response, however, is Italy's (17 percent).[35]

Given America's history as a British colony, it is particularly instructive to compare the findings from British and American polls. More than three times as many Britons as Americans say they do not

attend church. Does this correlate with an increased acceptance of evolutionary theory? Apparently yes: 48 percent of Britons embrace evolution as the best explanation for the development of life. It is always
tricky to compare results across polls, but the relevant comparison is
surely with the 13 percent of Americans who embrace evolution as
having occurred without divine intervention. Although the BBC presented the data from the British poll by trumpeting "Britons uncon-
vinced on evolution," it is equally true that Britons are far more
convinced than their counterparts across the ocean.[36]

Perhaps it is the case that Europe is filled with believers who wish
to relate to their God in more private ways than do the majority of
Americans. Though some Europeans are creationists or ID advocates,
there is no furious public, political, or legal battle in Europe over religion and science or what to teach in science classes. One British scholar
of religion, for instance, bluntly says that Europeans think George
Bush, with his constant, Bible-laced reference to his faith, "must be a religious nut."[37]

FROM COURT TO CAMP

Canoeing, horseback riding, and campfire ceremonies are staples of
kids' summer camps across the country. These activities go on at
Camp Quest in Ohio, too, but the goals here go beyond fun, self-
improvement, and bonding. Camp Quest's aim is to offer a safe haven
for the children of atheists and agnostics, a haven "beyond belief," as the
camp's slogan says. In the United States, children raised outside the
bounds of religion are statistical exceptions, and they pay the price for
being different. Campers report not only that they are called names
("devil worshiper") but also that they are pressured to turn toward belief in God by people who express concern for their souls.[38]

Some atheists respond to this intolerance in equal measure. Before
the Ethical Culture Society in New York, the science writer Natalie
Angier spoke on "raising healthy, 100 percent guaranteed god free children." Angier and her husband are atheists. When asked whether she

believes in God, their young daughter "crinkled up her nose at me," said Angier, "like I had mentioned something distasteful, like spinach and liver, or kissing a boy, and said, No!" The girl explained her reaction by saying (among other things) that "if somebody gets sick, I wouldn't just pray to god he or she gets better, I would try to buy some medicine for them, to help them get better." In her speech, Angier said, "Oh, I liked that answer. . . . This sounded to me like, what do you call it, a value system" and approved the fact that her daughter "likes to see things for herself before believing in them."[39]

It is not my usual practice to question a mother's pride in her young daughter. But when Angier made a public speech, she opened her remarks—not the little girl's—to public scrutiny. Can Angier possibly mean to imply that she, herself, needs to see a thing to believe in it? I doubt it, for I am certain that she "believes in" atoms, and far-distant stars, and other invisible glories of our universe, all known to scientists through hypothesis-testing and evidence-gathering despite their invisibility.

More troubling is Angier's simplistic equation of religion with prayer—and with prayer in the absence of a value system, at that, at least if you read her remarks a certain way. Theologians would remind Angier that compassionate action is the heart of religious expression. (That all religious people do not practice this ideal consistently does not change the ideal's importance, or, if you will, the ideal's reality.)

But Angier's superficial treatment of religion pales next to the outright hostility of well-known authors Daniel Dennett and Richard Dawkins. In his latest book, *Breaking the Spell*, Dennett explains religious behavior in terms of memes, idea-like bits that get transferred from person to person in cultural ways just like genes get passed along in biological ways. Memes survive when they outcompete other memes, a rivalry in which religion-oriented memes have clearly excelled throughout human history.

In describing their "capture" of a host, Dennett assigns an active role to memes.[40] As Dennett admits, this is basic evolutionary psychology, and closely related to Boyer's brain-based inference systems with a nod to Hamer's God gene. Two aspects of Dennett's views are worth closer scrutiny, however.

Dennett seriously misunderstands the nature of certain religious practices. Religious rituals, for instance, are for Dennett primarily "memory-enhancement processes" in aid of meme replication.[41] Any emotional component of religious expression is in service of the memes. Shamans engage in deception, not transformative encounter with the sacred. Consider what Dennett has to say about shamanic rituals that include, as I interpret the relevant passages, healing ceremonies as well as walking on hot coals: "One of the most interesting facts about these unmistakable acts of deceit is that the practitioners, when pressed by inquiring anthropologists, exhibit a range of responses. Sometimes we get a candid admission that they are knowingly using the tricks of stage magic to gull their clients . . . And sometimes, more interestingly, a sort of holy fog of incomprehension and mystery swiftly descends . . . These shamans are not quite con men—not all of them, at any rate—and yet they know that the effects they achieve are trade secrets that must not be revealed to the uninitiated for fear of diminishing their effects."[42] Anyone who has read *Black Elk Speaks* or other accounts of shamans' experiences may find, as I did, this passage to be breathtakingly ignorant.

Second, Dennett endeavors with evangelical zeal to persuade religious people of their folly. Those people who see Dennett as "just another liberal professor trying to cajole them out of some of their convictions" are, he admits, "dead right."[43] But this task is difficult, akin to an attempt "to persuade your friend with the cancer symptoms that she really ought to see a doctor *now* . . ."[44] Religious people stubbornly refuse to adopt a reasoned outlook: "So isn't my belief that belief in evolution is the path to salvation a religion? No; there is a major difference. We who love evolution do not honor those whose love of evolution prevents them from thinking clearly and rationally about it!"[45]

Richard Dawkins is a brilliant evolutionary thinker who also routinely opts to insult the faithful. Speaking at a meeting sponsored by the New York Institute for the Humanities, Dawkins sneered at those who would foster an interface between science and religion. Evolution, he said flatly, is antireligious. He did not stop there. Railing against the United States as a country of "religious maniacs," he yet reserved his most acerbic comments for scientists who embrace religion: "I think that scientists who say they are Catholics or Jews or Muslims owe it to us to say how they reconcile this with the sort of petty, cheap, parochial, niggling religion which goes with those titles."[46]

A large number of American scientists would have some explaining to do if Dawkins's wish were granted. About two-thirds of scientists from "elite research universities" say they believe in God.[47] This percentage varies a bit according to the scientists' field of study. Almost 38 percent of natural scientists (physicists, chemists, biologists, and others in related fields) say they do not believe in God, compared with 31 percent of social scientists. The philosopher Mary Midgely argues an irony: for some scientists, evolution *is* religion. Midgely says that when evolutionary scientists include in their writing "remarkable prophetic and metaphysical passages" and make "startling suggestions about vast themes such as immortality, human destiny, and the meaning of life," they embrace a religion.[48] It's beside the point that these very same scientists may say they do not believe in God; not all religions entail belief in any deity. The scientists who adopt evolutionism (Midgely's term) promote an ideology through "ambitious systems of thought, designed to articulate, defend and justify their ideas." These scientists, in short try to convert others.[49]

Midgely's points aside, that the majority of scientists believe in God strikes me as not very surprising. Here we arrive at what I take to be the singularly most toxic result of the poisonous divide: the notion that one must choose between believing in God and accepting evolutionary theory. One may so choose, of course. One may argue that the evolutionary evidence simply makes a supernatural explanation for life on earth unnecessary, or one may argue that evolution threatens the omnipotence of one's god. Many, many do argue this way, and some feel forced into arguing this way. Every year I encounter students who were raised, or educated, or both, to make a choice: God *or* evolution. These students sometimes find my classes quite painful because the material causes them to reconsider the choice they have made—or, I hope, to reconsider the need for a choice in the first place. What options are there for a person who accepts evolutionary theory and wants to think about God?[50]

CHOOSING GOD (OR NOT)

The evolutionary biologist Richard Dawkins is hostile to religion, as we have already seen and as is evident in so many of his utterances: "Science has no way to disprove the existence of a supreme being (this is strictly true). Therefore belief (or disbelief) in a supreme being is a matter of pure individual inclination, and they are therefore both equally deserving of respectful attention! When you say it like that the fallacy is almost self-evident: we hardly need spell out the *reductio ad absurdum*. To borrow a point from Bertrand Russell, we must be equally agnostic about the theory that there is a china teapot in elliptical orbit around the Sun."[51]

Dawkins's sneering has not gone unnoticed. The Roman Catholic theologian John Haught, in his book *Deeper Than Darwin*, notes the dismissiveness inherent in assuming that humans began to "imagine" God *in the sense that God is imaginary*. Haught sees three possibilities for understanding the "big questions" of life on earth—to wit, how life developed and how it changes over time.

In this first view, a person may accept the answers offered either by science or by religion, but not by both. The two realms are wholly incompatible. *Either* God (or gods or spirits) is real, and shapes our lives, *or* science is right and evolution shaped our world. Dawkins fits well into this category because for him, only one tenable choice exists for grown-ups who wish to engage with the real world. Science is about truth, and religion is about illusion.[52]

Also in Haught's first camp are theologians and believers who embrace divine providence to the complete exclusion of the theory of evolution. Notions of evolutionary change are as superfluous, in their view, for understanding the world as God is in the view of Dawkins.

In the second view, any forced choice between religion and science is unnecessary. Religion and science are complementary, but quite separate. No contradiction exists when a priest, rabbi, imam, or indeed any one of us embraces the wisdom one may find in the Bible, Torah, or Koran on a given morning, and finds equal wisdom in Darwin's writings that same night. A schema offered by the renowned Stephen Jay Gould, that of non-overlapping magisteria (NOMA), fits here. Gould thought that science and religion were best grasped as mutually respectful partners, the one (science) about knowledge and the other (religion) about values and ethics. For scientists, there is nothing to learn from religion about evolution. For religious people, there is nothing to learn from science about God.

For many religious thinkers as well as many scientists, NOMA is a workable idea, even a beautiful one. It allows for the peaceful coexistence of two equally glorious ways of moving through our world. But Haught sees NOMA a good deal less positively. Doesn't it deny to religion any link with objective reality, he asks? What happens to divine providence and transcendent reality with a NOMA framework? And doesn't NOMA deny to science any way to speak about the divine?

Haught yearns for a third way. He believes no one needs to choose between religion and science (though, of course, anyone may so choose). He believes no one needs to divorce God from knowledge and reality, and no one needs to fear that evolution can add nothing to divine accounts of life in the universe. Haught's approach calls for a deeper reading of the universe from an evolutionary perspective and from religious perspectives alike: "Above all," Haught says, "evolution requires a revolution in our thoughts about God."[53]

For Haught, evolutionary science allows us to grasp a fundamental truth about the nature of the universe: even with all its flaws, the universe is exactly consonant with what a truly loving God would create. That we humans cause suffering, create unspeakable violence, and thus fail each other in terrible ways, is not evidence for God's absence. These aspects of humanity exist side by side with compassion, love, and transforming action. They do so because a loving, noncoercive God made a world that itself is still coming into being. God affords opportunities and experimentation, not a complete and perfected creation. The notion of an unfinished universe, inherited from evolutionary biology, allows us to understand God in this way.

What a range of ideas, from Dawkins's scorn for religion, to Gould's embrace of both religion and science that still splits religion from science, to Haught's notion that evolution helps us understand more deeply the actions of a loving God.

Haught's openness to bringing religion and science closer together resonates with me. Science is as much about belongingness as it is about genes, as we have seen, and belongingness is fundamental to religion. A strong dissatisfaction with the gene- and brain-centered scenarios—the belief that *science can do better* in exploring the origins of religion—motivated me to write this book. Science can look head-on at humanity's hunger for the sacred, a hunger that is far more than a mere offshoot of the workings of our genes or brains and far more than an illusion akin to a Chinese teapot orbiting the sun.

FINAL THOUGHTS

We finish, then, where we started, with Buber's notion that human relating to God is "wrapped in a cloud but reveals itself, it lacks but creates language." Agnosis, not-knowing, exists in beautiful tension with the striving to know ever more and ever deeper. Religion is based on faith but may include a deep desire to better know God (or gods). Science is based on knowing but may include an acceptance of not-knowing as a condition of being human.[54]

I do not believe that science can "explain" religion. An evolutionary perspective will probably never be able to pinpoint the reasons why an apelike creature capable of empathy and meaning-making developed into a species that sings the praises of God and shakes in fear at the wrath of the gods, who goes to war in the name of religion and who sacrifices all worldly comforts in order to honor God by doing good for others.

I do believe that science can explain something meaningful about the evolution of the religious imagination. The religious impulse is rooted in a deep longing for the emotional meaning-making with other beings that is so fundamental to the prehistory of our species. We crave belongingness, and we seek it with other people, with other animals, and with spirits, gods, and God, on earth and in unearthly realms.

The story that has been told in these pages is one of engagement with the mysteries all around us, the mysteries that humans have for so very long experienced in ways seamlessly bound up with emotional meaning-making, with creativity in art and dance, and with emotional mutuality among social partners. For millions of years, human ancestors derived meaning from mutuality; late in our prehistory, humans began to seek belongingness in the sacred realm as well as in the daily rhythms of small-group life. Emotions that we had been content, before, to create with those we could see, hear, and touch, we now began to create in relation with sacred beings.

That we evolved as spiritual creatures because we evolved first as meaning-makers in emotional relation with each other is a message grounded in the evolutionary perspective, and in hope. When belongingness runs amok, it can become xenophobia, and people may begin to act out of fear and hatred of others. Yet the power of belongingness amounts to the power to base our lives, the lives of all humans who are intertwined in a globe-sized web of belongingness, on an understanding that we all come from the same roots. We evolved as primates; as ancestral hominids in Africa; as settlers of all the corners of our world; and finally as people who act in relation to sacred beings. We are primates still, able to embrace the expression of different faiths, or no faith at all, as we continue to make meaning through belongingness.

ACKNOWLEDGMENTS

MY PRIMARY THANKS go to Trace Murphy at Doubleday Religion, who first proposed that I write about the prehistory of religion. Trace's instinct for how best to present an evolutionary account made this a far better book than it otherwise would have been. I am grateful as well to Doubleday for Jolanta Benal's astute copyediting and Darya Porat's capable assistance in readying the manuscript.

The Faculty Research Program at the College of William & Mary made this book possible in another sense. The award of a year's freedom to read and write without interruption made all the difference. I thank Geoffrey Feiss and Carl Strikwerda at the College for years of support for my research.

The insights of John Haught and Stuart Shanker, who read the entire manuscript (in a slightly different version), added immeasurably to the final product. Through patient explanations of Neolithic archaeology, and a dash of editing, Mary Voigt improved the start of Chapter 6. I am fortunate to have these generous colleagues.

My research and writing on African apes has been supported financially by the College of William and Mary; the Guggenheim Foundation; the Harris Steel Foundation; the Language Research Center of Georgia State University; the Templeton Foundation; and the Wenner-Gren Foundation for Anthropological Research. I am particularly grateful to the National Zoological Park, Washington, D.C., for allowing me to conduct longitudinal research on a family of gorillas.

To Cindy Baker, Robert Bednarik, Alan Fogel, and Frans de Waal, my appreciation for their generous sharing of photographs.

To Jessa Crispin, thanks for making it worthwhile for me to look my mother in the eye and say, "I write for Bookslut."

To Mark Spahr, a magilla-sized thanks for a critical rewrite on a single paragraph and for more creative discussions than I can count.

For discussions about religion, science, and/or writing, and for unflagging support, I thank Joanne Bowen, Cahlene Cramer, Karen Flowe, Ron Flowe, Charles Hogg, Willow Powers, Stuart Shanker, and Joanne Tanner, and in a special longevity award, Stephen Wood.

For all those years of reading to me, and for a complete inability to refuse my requests in any bookstore on the planet, I thank Elizabeth King. For all their patience and love, I am beyond grateful, now and forever, to Charles Hogg and Sarah Hogg, plus of course to our four domestic cats, many feral cats, and one small rabbit.

NOTES

ONE: *Apes to Angels*

1. S. Mithen, *The Singing Neanderthals: The Origins of Music, Language, Mind, and Body* (London: Weidenfeld & Nicholson, 2005), 221.
2. Martin Buber, *I and Thou* (New York: Touchstone, 1970), 57.
3. Ibid., 62, 78, 80.
4. Christophe Boesch and H. Boesch-Achermann, *The Chimpanzee of the Tai Forest* (Oxford: Oxford University Press, 2000).
5. All information about the Fang, from here on, taken from Boyer, P., *Religion Explained* (New York: Basic Books, 2001).
6. Ibid., 141.
7. R. Coles, *The Secular Mind* (Princeton, N.J.: Princeton University Press, 1999), 32–34.
8. Boyer, *Religion Explained*, 297.
9. The choices I made in that research, noted in my first book, *The Information Continuum*, reflect a theoretical framework different from the one I now employ, as in my more recent book *The Dynamic Dance*.
10. Though it is, of course, possible to consider faith as different from belief, I will treat them as the same here.
11. Boyer, *Religion Explained*, 7–10.
12. Clifford Geertz, *The Interpretation of Cultures* (New York: Basic Books), 90.

13. Karen Armstrong, *The Spiral Staircase: My Climb Out of Darkness* (New York: Knopf, 2004).

14. Ibid., 42.

15. Ibid., 235–36.

16. Ibid., 265.

17. Ibid., 270.

18. Ibid., 302–3.

19. *The Encyclopedia of North American Indians* (Houghton Mifflin); available at http://college.hmco.com/history/readerscomp/naind/html/na_032600_religion.htm.

20. Quoted in Ursula Goodenough, *The Sacred Depths of Nature* (Oxford: Oxford University Press, 1998), 87.

21. Quoted in Patricia Leigh Brown, "A Native Spirit, Inside the Beltway," *The Washington Post*, September 9, 2004.

22. Ibid.

23. See S. F. Waugaman and D. Moretti-Langholtz, *We're Still Here* (Richmond, Va.: Palari Publishing, 2000).

24. Brian Hayden, *Shamans, Sorcerers and Saints* (Washington, D.C.: Smithsonian Books, 2003), 9.

25. from the New King James Version

26. Richard Bauckham, *Loving our fellow creatures*. The Anglican Society for the Welfare of Animals; available at http://www.aswa.org.uk/Articles/lovingour fellowc.html

27. I am grateful to Karen Flowe for inspiration in linking the arts and the study of religion.

TWO: *Imagining Apes*

1. Frans de Waal, *My Family Album* (Berkeley, Calif.: University of California Press, 2003).

2. Jane Goodall, "A Sorrow Beyond Tears," In M. Bekoff, ed., *The Smile of a Dolphin* (New York: Discovery Books, 2000), 140.

3. Masserman et al. (1964), reviewed in S. D. Preston and F.B.M. de Waal. "Empathy: Its Ultimate and Proximate Bases," *Behavioral and Brain Science* 25 (2002): 1–72.

4. Ibid.

5. Special thanks to Melanie Bond, biologist recently retired from the National Zoo.

6. D. Fouts and R. Fouts, "Our Emotional Kin." In Bekoff, *The Smile of a Dolphin*, 207.

7. See especially Chapter 2 of *The Dynamic Dance* (Cambridge, Mass.: Harvard University Press, 2004).

8. E. Ingmanson, "Empathy in a Bonobo." In R.W. Mitchell, ed., *Pretending and Imagination in Animals and Children* (Cambridge, Eng.: Cambridge University Press, 2002), 280–84.

9. Christophe Boesch and H. Boesch-Achermann, *The Chimpanzee of the Tai Forest* (Oxford: Oxford University Press, 2000).

10. Frans de Waal, *Good Natured* (Cambridge, Mass.: Harvard University Press, 1996), 210.

11. Alan Fogel, *Developing Through Relationships* (Chicago: University of Chicago Press, 1993).

12. But see Charles Goodwin, The interactive construction of a sentence in natural conversation. In G. Pathas, ed. *Everyday Language* (New York: Irvington Publishers, 1979, 97–121); Talbot Taylor, *Theorizing Language* (Amsterdam: Pergamon, 1997); and David F. Armstrong and Sherman L. Wilcox, *The Gestural Origins of Language* (Oxford: Oxford University Press, in press).

13. Fogel, *Developing Through Relationships*.

14. This insight was crucial to the work of the pioneering psychiatrist Murray Bowen; see http://www.thebowencenter.org/.

15. Stanley I. Greenspan and Stuart G. Shanker, *The First Idea* (New York: Da Capo Press, 2004).

16. See *The Dynamic Dance* for examples.

17. E. S. Savage-Rumbaugh and R. Lewin, *Kanzi* (New York: Wiley, 1994).

18. S. Kuroda, "Rocking Gesture as Communicative Behavior in the Wild Pygmy Chimpanzees in Wamba, Central Zaire," *Journal of Ethology* 2 (1984):127–37; quoted material appears on p. 135.

19. K. Armstrong, *A Short History of Myth* (Edinburgh: Canongate, 2005), 19.

20. Richley Crapo, *Anthropology of Religion* (New York: McGraw Hill, 2002).

21. C. S. Alcorta and R. Sosis, "Ritual, Emotion, and Sacred Symbols." *Human Nature* 16(4):323–359, 2005.

22. De Waal, *Good Natured*, 91–92.

23. Greenspan and Shanker, *The First Idea*, 126

24. J. Flack and F. de Waal, "Play Signaling and the Perception of Social Rules by Juvenile Chimpanzees," *Journal of Comparative Psychology* 118 (2004): 149–59.

25. John M. Watanabe and Barbara B. Smuts, "Cooperation, Commmitment, and Communications in the Evolution of Human Sociality." In Robert W. Sussman and Audrey R. Chapman, *The Origins and Nature of Sociality* (New York: Aldine, 2004), 288–309.

26. R. W. Mitchell, *Pretending and Imagination in Animals and Children* (Cambridge: Cambridge University Press, 2002), quoting from Keith and Cathy Hayes, *The Ape in Our House* (New York: Harper & Bros., 1951), 80–84.

27. R. Wrangham, "Making a Baby." In Bekoff, *The Smile of the Dolphin*, 27.

28. M. Donald, *A Mind So Rare* (New York: Norton, 2001), 26.

29. William Mullen, "One by One, Gorillas Pay Their Last Respects," *Chicago Tribune*, November 8, 2004.

THREE: *African Origins*

1. *A Short History of Nearly Everything* (New York: Broadway, 2003), 415.
2. M. Small, *Our Babies, Ourselves* (New York: Anchor Books, 1998), 7 (citing R. D. Martin).
3. Like almost everything else in biological anthropology, this taxonomy is not accepted by all, but it represents a fair consensus.
4. B. Wood, "Hominid Revelations from Chad," *Nature* 418 (July 11, 2002), 133–34; quoted material is on p. 133.
5. http://www.bbc.co.uk/science/cavemen/indepth/indepth1.shtml.
6. One is my course for the Teaching Company, which includes recommended readings; see "Biological Anthropology," at www.teachco.com.
7. G. Conroy, *Reconstructing Human Origins* (New York: Norton, 2005), 4.
8. S. Semaw et al., "Early Pliocene Hominids from Gona, Ethiopia," *Nature* 433 (2005): 301–305.
9. Conroy, *Reconstructing*, 146.
10. http://www.mnh.si.edu/anthro/humanorigins/ha/afri.html.
11. L. R. Backwell and P. d'Errico, "Evidence of Termite Foraging by Swartkrans Early Hominids," *Proceedings of the National Academy of Sciences* 98 (4) (2001): 1358–63.
12. Dean Falk, "Prelinguistic evolution in early hominids: Whence motherese?" *Behavioral and Brain Sciences* 27 (2004): 491–503.
13. Bednarik prefers this term to "archaeologist."
14. R. Bednarik, "The 'Australopithecine Cobble' from Makapansgat, South Africa," *South African Archaeological Bulletin* 53 (1993): 4–8; quoted material is on p. 5.
15. Bednarik prefers the term "hominins" to my "hominids" to refer to the creatures I am describing: we both refer to human ancestors who lived after the split with the great apes.
16. Brian Hayden, *Shamans, Sorcerers and Saints* (Washington, D.C.: Smithsonian Books, 2003), 95.
17. S. Semaw, et al. "The World's Oldest Stone Artifacts from Gona, Ethiopia," *Journal of Archaeological Science* 27 (2000): 1197–1214; quoted material is on p. 1211.
18. A. Whiten et al. "Cultures in Chimpanzees," *Nature* 399 (1999): 682–85.
19. W. McGrew, *The Cultured Chimpanzee* (Cambridge, Eng.: Cambridge University Press, 2004), 125.
20. Here and in the following two paragraphs I rely on M. Panger et al., "Older Than the Oldowan? Rethinking the Emergence of Hominin Tool Use," *Evolutionary Anthropology* 11 (2002): 235–45.
21. M. R. Leary and N. R. Buttermore, "The Evolution of the Human Self: Tracing the Natural History of Self-awareness," *Journal of the Theory of Social Behavior*, 33:4 (2003): 365–404; quoted material is on p. 376.
22. S. Tarlow, "Emotion in Archaeology," *Current Anthropology* 41 (5) (2000): 713–46; quoted material is on p. 719.
23. Stanley I. Greenspan and Stuart G. Shanker, *The First Idea* (New York: Da Capo Press, 2004).

24. Ibid.

25. Hayden, *Shamans, Sorcerers and Saints*, 97.

26. Ibid., 99.

27. I. Wunn, "Beginnings of Religion," *Numen* 47 (2000): 417–52; quoted material is on pp. 447, 448.

28. Hayden, *Shamans, Sorcerers and Saints*, 95.

29. In *A Short History of Myth* (New York: Canongate, 2005), Karen Armstrong makes this same point on p. 25.

30. Falk, "Prelinguistic evolution in early hominids: Whence motherese?" *Behavioral and Brain Sciences* 27 (2004): 491–503.

FOUR: *Cave Stories: Neandertals*

1. Brian Hayden, *Shamans, Sorcerers and Saints*. (Washington, D.C.: Smithsonian Books, 2003), 112.

2. Ibid., 114.

3. Ibid., 117.

4. Philip and Carol Zaleski, *Prayer* (Boston: Houghton Mifflin, 2005). See p. 17.

5. Stanley I. Greenspan and Stuart G. Shanker, *The First Idea* (New York: Da Capo Press, 2004).

6. Steven Mithen, *The Singing Neanderthal* (London: Weidenfeld and Nicolson, 2005).

7. R. Rappaport, *Ritual and Religion in the Making of Humanity* (Cambridge, Eng.: Cambridge University Press, 1999), 24.

8. A. Gibbons, "Chinese Stone Tools Reveal High-Tech *Homo erectus*," *Science* 287 (2000): 1566.

9. T. White, "Cut Marks on the Bodo Cranium," *American Journal of Physical Anthropology* 69 (1986): 503–509.

10. Hayden, *Shamans, Sorcerers and Saints*, 97.

11. R. Bednarik, "A Figurine from the African Acheulian," *Current Anthropology* 44(3) (2003): 405–413.

12. Steven Mithen, *The Singing Neanderthal* (London: Weidenfeld and Nicolson, 2005, 218), evaluating the research of the archaeologist J. L. Arsugua and his colleagues.

13. Hayden, *Shamans, Sorcerers and Saints*, 98.

14. Ian Tattersall, "Once We Were Not Alone," in *Scientific American's* special issue, New Look at Human Evolution, 20–27; quoted material is on p. 25.

15. João Zilhão and Francesco d'Errico, "A Case for Neandertal Culture," in *Scientific American's* special issue New Look at Human Evolution, 34–35; quoted material is on p. 35.

16. BBC interview, online at http://news.bbc.co.uk/2/hi/science/nature/3256228.stm.

17. Steven Mithen, *The Singing Neanderthal* (London: Weidenfeld and Nicolson, 2005), 245.

18. Ibid., 244.

19. Hayden, *Shamans, Sorcerers and Saints*, 117.
20. Hayden, *Shamans, Sorcerers and Saints*, 102.
21. But for a different view, see D. Lewis-Williams, *The Mind in the Cave* (London: Thames & Hudson, 2003).
22. Zaleski and Zaleski, *Prayer*, 5.
23. A. Defleur et al., "Neanderthal Cannibalism at Moula-Guercy, Ardeche, France," *Science* 286 (1999): 128–31.
24. Tim D. White, "Once Were Cannibals," in *Scientific American's* special issue New Look at Human Evolution, 86–93.
25. http://news.bbc.co.uk/1/low/sci/tech/462048.stm.

FIVE: *More Cave Stories: Homo Sapiens*

1. http://www.talkorigins.org/faqs/cosmo.html.
2. Peter Brown et al., "A New Small-Bodied Hominid from the Late Pleistocene of Flores, Indonesia," *Nature* 431 (2004): 1055–1061.
3. K. Wong, "The Littlest Human," *Scientific American*, February 2005, 61.
4. V. Formicola and A. P. Buzhilova, "Double Child Burial from Sunghir (Russia)," *American Journal of Physical Anthropology* 124 (2004): 189–98.
5. R. White, posted on Institute for Ice Age Studies website, http://www.insticeagestudies.com/.
6. Support for this comes from D. Lewis-Williams, *The Mind in the Cave* (London: Thames & Hudson, 2002), 80.
7. See Formicola and Buzhilova, "Double Child Burial," for an explanation based on anatomical pathology.
8. Some color photos appear in Lewis-Williams, *Mind in the Cave*; a good virtual tour can be found at http://www.culture.gouv.fr/culture/arcnat/ lascaux/en/.
9. D. Lewis-Williams, *The Mind in the Cave* (London: Thames & Hudson, 2002), 44, 237.
10. Ibid., 249–50.
11. Ibid., 44.
12. Unless otherwise noted, my understanding of shamanism relies on Lewis-Williams' *The Mind in the Cave* (London: Thames & Hudson, 2002) and Brian Hayden's *Shamans, Sorcerers and Saints* (Washington, D.C.: Smithsonian Books, 2003), plus information made available by Harner at http://www.shamanism.org.
13. http://www.shamanism.org/.
14. J. G. Neihardt, *Black Elk Speaks* (Lincoln: University of Nebraska Press, 2004).
15. Ibid., 37.
16. Ibid., 154.
17. Hayden, *Shamans, Sorcerers and Saints*.
18. http://news.bbc.co.uk/1/hi/sci/tech/871930.stm.
19. http://news.bbc.co.uk/1/hi/sci/tech/3142488.stm.
20. Estimate by Gladkih, Kornietz and Soffer, cited in Hayden, *Shamans, Sorcerers and Saints*, 159.

21. Hayden, *Shamans, Sorcerers and Saints*, based on the work of Vandiver, Soffer, Klima, and Svoboda.

22. Boston: Beacon Press, 2000.

23. O. Soffer et al., "The 'Venus' Figurines," *Current Anthropology* 41(4) (2000): 511–37; quoted material is on p. 524.

24. Hayden, *Shamans, Sorcerers and Saints*, 129.

25. P. Mellars, "The Impossible Coincidence: A Single-Species Model for the Origin of Modern Human Behavior in Europe," *Evolutionary Anthropology* 14 (2005): 12–27.

26. http://news.bbc.co.uk/1/hi/sci/tech/1753326.stm.

27. This is the work of Stanley Ambrose, recounted in R. Klein, *Dawn of Human Culture* (New York: Wiley, 2002).

28. http://www.natmus.cul.na/events/rockart/open.html.

29. http://news.bbc.co.uk/1/hi/sci/tech/3310233.stm.

30. http://news.nationalgeographic.com/news/2003/02/0224_030224_mungoman.html.

31. S. McBrearty and A. S. Brooks, "The Revolution That Wasn't," *Journal of Human Evolution* 39(5) (2000): 453–563.

32. http://dsc.discovery.com/news/briefs/20050502/neanderthal.html.

SIX: *Transformations in Time*

1. O. Soffer et al., "The 'Venus' Figurines," *Current Anthropology* 41 (4): 511–37; quoted material is on p. 512.

2. M. Balter, "Seeds of Civilization," *Smithsonian* May 2005.

3. M. Balter, *The Goddess and the Bull. Catalhöyük: An Archaeological Journey to the Dawn of Civilization* (New York: Free Press, 2005).

4. http://catal.arch.cam.ac.uk/catal/index.html.

5. Balter, *The Goddess and the Bull*.

6. Her article by this title appears in *Life in Neolithic Farming Communities*, ed. I. Kuijt (New York: Kluwer/Plenum, 2000), 253–93.

7. Ibid.

8. New York: Knopf, 2006, 395–96.

9. Ibid., 208.

10. Ibid., 256.

11. Ibid., 379.

12. Ibid., xiii.

13. Ibid., 399.

14. W. A. Watkins et al., "Twelve Years of Tracking 52-Hz Whale Calls from a Unique Source in the North Pacific," *Deep Sea Research* (Dec. 2004), 1889–1901.

15. Kate Stafford quoted in Andrew C. Revkin, "A Song of Solitude," *New York Times*, Dec. 26, 2004.

16. H. Queiroz and A. E. Magurran, "Safety in Numbers? Shoaling Behaviour of the Amazonian Red-bellied Piranha," *Biology Letters* (May 10, 2005).

17. J. Silk, S. C. Alberts, J. Altmann. "Social Bonds of Female Baboons Enhance Infant Survival," *Science* 302 (November 14, 2003): 1231–34.

18. R. F. Baumeister and M. R. Leary. "The Need to Belong: Desire for Interpersonal Attachments as a Fundamental Human Motivation," *Psychological Bulletin* 117(3) (1995): 497–529.

19. Ibid., 515.

20. Gustave Flaubert, *Madame Bovary*, Geoffrey Wall, trans. (London: Penguin, 2003), 217–18.

21. T. Singer, et al. "Empathy for Pain Involves the Affective but Not Sensory Components of Pain," *Science* 303 (2004): 1157–62; C. Holden, "Imaging Studies Show How Brain Thinks About Pain," *Science* 303 (2004): 1121.

22. G. S. Berns et al., "Neurobiological Correlates of Social Conformity and Independence During Mental Rotation," *Biological Psychiatry* 58 (3): 245–253.

23. But see the first chapter of *The Dynamic Dance*.

24. Thanks to Springsteen fans who replied to my query, posted in January 2006, to the Underworld forum at http://www.greasylake.org/.

25. http://www.csuchico.edu/pub/inside/archive/01_09_13/spiritualpop.html.

26. R. Stark, *The Rise of Christianity* (San Francisco: HarperCollins, 1997), 19.

27. Ibid., 161.

28. Tanya M. Luhrmann, "Metakinesis: How God Becomes Intimate in Contemporary U.S. Christianity," *American Anthropologist* 106(3): 518–28, 2004.

29. Stark, *Rise of Christianity*, 86.

30. Turnbull's research described in Richley Crapo, *Anthropology of Religion* (New York: McGraw Hill, 2002).

31. *The Great Transformation* (New York: Knopf, 2006), 330.

32. Emile Durkheim, *The Elementary Forms of the Religious Life*. Carol Cosman, trans. (Oxford: Oxford University Press, 2001), 311.

33. Ibid., 318.

34. Ellen Dissanayake, *Art and Intimacy* (Seattle: University of Washington Press, 2002), 78, quoting anthropologist David Guss.

35. Ibid., 64.

36. See also Steven Mithen, *The Singing Neanderthal* (London: Weidenfeld and Nicolson, 2005).

37. Dissanayake, *Art and Intimacy*, 204.

38. Ibid., 139.

39. See http://www.councilhd.ca.

40. Stanley I. Greenspan and Stuart G. Shanker, *The First Idea* (New York: Da Capo Press, 2004), 2.

41. Ibid., 28.

42. The example is mine; the quotation is from ibid., 70.

43. Ibid., 31–32; I have replaced the word "baby" with the word "child" and changed the child's gender throughout.

44. Ibid., 157.

45. Karen Armstrong, *A History of God* (New York: Ballantine, 1993), 397.

SEVEN: *Is God in the Genes?*

1. New York: Doubleday, 2004.
2. *Time,* October 25, 2004.
3. My work with the Council of Human Development, and editing the *Journal of Developmental Processes,* is in large part a response to gene obsession. See www.councilhd.ca.
4. C. Zimmer, "Faith-Boosting Genes," *Scientific American,* October 2004, 111.
5. Hamer, *The God Gene,* 18.
6. Ibid., 53.
7. Evelyn Fox Keller, *The Century of the Gene* (Cambridge, Mass.: Harvard University Press, 2000), 71.
8. Jerry Coyne, "The Gene Is Dead; Long Live the Gene," *Nature* 408 (2000): 26–27.
9. Stanley I. Greenspan and Stuart G. Shanker, *The First Idea* (New York: Da Capo Press, 2004), 286.
10. Pascal Boyer, *Religion Explained* (New York: Basic Books, 2001), 128.
11. Ibid., 330.
12. John Tooby and Leda Cosmides, "The Psychological Foundations of Culture," in *The Adapted Mind: Evolutionary Psychology and the Generation of Culture,* J. H. Barkow, L. Cosmides, and J. Tooby, eds.) Oxford: Oxford University Press, 1992), 113.
13. Boyer, *Religion Explained,* 234.
14. Ibid., 161.
15. Ibid.
16. See *Attachment, Evolution and the Psychology of Religion* (New York: The Guilford Press, 2005), by my William & Mary colleague Lee Kirkpatrick; see also S. Atran, *In Gods We Trust* (Oxford University Press, 2002).
17. M. Holloway, "The Mutable Brain," *Scientific American,* Sept. 2003, 73–85.
18. S. Blakeslee, "Exercising Toward Repair of the Spinal Cord," *New York Times,* Sept. 22, 2002.
19. Ibid.
20. Both studies reviewed by N. Weinberger, "Music and the Brain," *Scientific American,* November 2004, 88–95.
21. Boyer, *Religion Explained,* 198.
22. New York: Da Capo Press.
23. Thanks to Mark Spahr for the phrase about a species-wide attack.
24. Boyer, *Religion Explained,* p. 20.
25. D. S. Wilson's *Darwin's Cathedral* (Chicago: University of Chicago Press, 2002) is more socially oriented than the other theories.
26. Stephen Jay Gould's *The Mismeasure of Man* (New York: W. W. Norton & Co., revised ed. 1996, originally published 1981) is a superb recounting of this phenomenon.
27. Thanks to Tom Givon for his permission to quote the posting.

EIGHT: *God and Science in Twenty-first-Century America*

1. Thanks to my capable assistant, Sarah Elizabeth Hogg.
2. B. Phillips, "Religion Journal: On God, with Roller Coasters Nearby." *New York Times,* June 28, 2003.
3. A. Goodenough, "Darwin-free Fun for Creationists," *New York Times,* May 1, 2004.
4. J. Leland, "A Church That Packs Them In, 16,000 at a Time," *New York Times,* July 18, 2005.
5. C. DeLaFuente, "The Call to the Torah, Now Heeded Online," *New York Times,* July 1, 2004.
6. B. J. Feder, "Services at the First Church of Cyberspace," *New York Times,* May 15, 2004.
7. The examples in this paragraph are drawn from D. D. Kirkpatrick, "What They Said: What Was Heard: Speaking in the Tongue of Evangelicals," *New York Times,* Oct. 17, 2004.
8. Quoted in K. Connie Kang, "Feeling God's Spirit Through the Arts," *Washington Post,* Feb. 5, 2005.
9. Quoted in Martha Ullman West, "Moved by the Spirit: Religion's Role in Concert Dance," *Chronicle of Higher Education,* July 22, 2005.
10. Judith Jamison quoted in ibid.
11. This first set of numbers from a Harris Poll of 2,201 adults conducted in 2003.
12. Angel statistics from Scripps Howard News Service–Ohio University poll of 1,127 adults in 2001.
13. Based on an ABC News/Primetime poll of 1,011 adults in February 2004.
14. Based on a CBS news poll of 885 adults in November 2004.
15. Ibid.
16. "Afraid to Discuss Evolution" (editorial), *New York Times,* Feb. 4, 2005.
17. C. Dean, "Scientific Savvy? In U.S., Not Much," *New York Times,* Aug. 30, 2005.
18. Thanks to Charles Hogg for this point.
19. http://www.answersingenesis.org/creation/v24/i4/canyon.asp.
20. http://www.answersingenesis.org/home/area/re1/chapter3.asp.
21. R. Dawkins, "Where D'you Get Those Peepers?" *New Statesman and Society,* June 16, 1995.
22. For more information, see *www.natcenscied.org.*
23. *Time,* Aug. 15, 2005.
24. The quotations are from J. Neuman, "Inspiration for Doubters of Darwin," *Los Angeles Times,* Aug. 3, 2005.
25. Buttars, D. Chris, "Evolution lacks fossil link," *USA Today,* Aug. 8, 2005.
26. Neuman, "Inspiration for Doubters of Darwin," *Los Angeles Times,* Aug. 20, 2005; J. Garofoli, "Bush Pushes Very Hot Button," *San Francisco Chronicle,* Aug. 8, 2005.
27. *Atlanta Journal-Constitution,* Aug. 7, 2005.
28. http://www.nsta.org/main/news/stories/nsta_story.php?news_story_ID=50792.
29. http://www.discovery.org/csc/topQuestions.php.
30. See http://www.discovery.org/csc/freeSpeechEvolCampMain.php.

31. J. Wilgoren, "Politicized Scholars Put Evolution on the Defensive," *New York Times*, Aug. 21, 2005.

32. Ibid.

33. The findings were based on interviews with 8,600 adults. See http://across.co.nz/WhatEuropeansBelieve.html.

34. F. Bruni, "Mainline Christianity Withering in Europe," *New York Times*, Oct. 13, 2003.

35. N. Knox, "Religion Takes a Back Seat in Western Europe," *USA Today*, Aug. 10, 2005.

36. http://news.bbc.co.uk/1/hi/sci/tech/4648598.stm.

37. Philip Jenkins quoted in Bruni, "Mainline Christianity Withering"; C. Krauss, "In God We Trust . . . Canadians Aren't So Sure," *New York Times*, March 26, 2003.

38. S. Hansen, "Summer Camp That's a Piece of Heaven for the Children, but Please, No Worshiping," *New York Times*, June 29, 2005.

39. http://www.cfimetrony.org/transcripts_html/natalie_angier.html.

40. Daniel C. Dennett, *Breaking the Spell* (New York: Viking, 2006), 186.

41. Ibid., 142.

42. Ibid., 166.

43. Ibid., 53.

44. Ibid., 39.

45. Ibid., 268.

46. Quoted in S. S. Hall, "Darwin's Rottweiler," *Discover*, Sept. 2005.

47. http://www.livescience.com/othernews/050811_scientists_god.html.

48. Mary Midgely, *Evolution as a Religion* (London: Routledge, 2002), viii–ix.

49. Ibid., 17–18.

50. As this book went to press, I found Francis Collins' book *The Language of God* (New York: Free Press, 2006) to be an excellent response to this question.

51. R. Dawkins, *A Devil's Chaplain* (Chicago: University of Chicago Press, 2003), 149.

52. Ibid., 160.

53. J. Haught, *Deeper Than Darwin* (Boulder, Colo.: Westview, 2003), 162.

54. A. Fogel, "Dynamic Systems Research on Interindividual Communication: The Transformation of Meaning-making," *Journal of Developmental Processes*, 1(1), 7–30.

INDEX